Stress and Poverty

Michael Breitenbach • Elisabeth Kapferer
Clemens Sedmak

Stress and Poverty

A Cross-Disciplinary Investigation of Stress in Cells, Individuals, and Society

 Springer

Michael Breitenbach
Department of Cell Biology
University of Salzburg
Salzburg, Austria

Elisabeth Kapferer
Centre for Ethics & Poverty Research
University of Salzburg
Salzburg, Austria

Clemens Sedmak
Keough School of Global Affairs
University of Notre Dame
Notre Dame, IN, USA

ISBN 978-3-030-77740-1   ISBN 978-3-030-77738-8   (eBook)
https://doi.org/10.1007/978-3-030-77738-8

This Springer imprint is published by the registered company Springer Nature Switzerland AG.
The registered company address is: Gewerbestrasse 11, 6330 Cham, Switzerland

Preface

Stress is a biological phenomenon that can be described in biological terms, even at a cellular level. Poverty is a social (and societal) challenge that can be examined using both the descriptive language of the social sciences and the normative language of moral philosophy and ethics. Poverty is experienced as stressful, and many studies discuss the link between poverty and stress.

A book about stress and poverty connects a biological phenomenon with a social and societal challenge. This book is about poverty "and" stress, not in the sense of an addition—we talk about stress, and then we talk about poverty—but in the sense of a connection in which we talk about poverty insofar as it is related to stress, and we talk about stress insofar as it is related to poverty. By exploring this link, we are exploring an intersection, the intersection between stress and poverty. How can stress research help us to reach a deeper understanding of poverty, and how can poverty research contribute to a richer discourse on stress?

The main claims of the book are threefold: Stress research and poverty research are mutually relevant conversation partners, and these cross-disciplinary interactions elevate our understanding. Even though the language may be different, the term "stress," used in selected biological fields and in the various fields of poverty research, refers to the same series of phenomena. Finally, the intersection of stress and poverty is not only of theoretical interest but also politically relevant because of the possibility of policy interventions that are explicitly stress-research sensitive.

This book is the result of many conversations and also the reflections of a small but intense learning community, a circle of three colleagues talking to and learning from each other (a process, which also led to the glossary at the end of the book[1]). Michael Breitenbach is a biochemist working in the field of molecular genetics; Elisabeth Kapferer is a literature scholar with a special focus on social exclusion and the discourse on poverty; and Clemens Sedmak is an ethicist with a special interest in poverty research. Even though the main responsibility for the chapters of this book (and the main work on these chapters) has been carefully divided, the text of this book is designed to be one coherent piece of research and writing.

Breitenbach is the main author of Chaps. 2, 3, 4, and 5. Chapter 6 is the product of a close collaboration between Breitenbach and Kapferer. She took the lead in writing the introduction (Chap. 1), in bringing the chapters together, and in building bridges between the biological discourse, the social science discourse, and the normative deliberations. Kapferer and Sedmak authored Chaps. 8 and 9, and Sedmak is responsible for Chaps. 7, 11, and 13. Chapter 10 was in large part written by Sedmak, with some smaller parts contributed by Kapferer; Chapter 12 was also mainly written by Sedmak, with a short biological subsection authored by Breitenbach.

The structure of the book is intentional in that the biological foundations are presented first, before we move into the social science discourse. We have worked hard to bring together our disparate disciplines in a coherent book.

We are grateful to the publishing house for its assistance and flexibility, with special thanks to our editor Tanja Weyandt for her support.

In lieu of a dedication, we would like to say: In writing this book, we have kept in our minds and hearts the many people struggling with stressful lives and the consequences of social exclusion. In some small way, maybe in teaching and training settings, we hope that this book can contribute to a deeper awareness of these challenges many members of the human family face in their daily lives.

Salzburg, Austria   Michael Breitenbach
Salzburg, Austria   Elisabeth Kapferer
Notre Dame, IN   Clemens Sedmak

---

[1] Glossary terms are bolded at first mention in each chapter.

Acknowledgments

We want to express our gratitude to the following colleagues for their invaluable help:

To Ian Dawes, for critically reading the "biological" chapters (Chaps. 2–6) and for giving us many ideas about the biology of stress;

To Bernhard Iglseder, for his help with all our questions concerning the human brain and for critically reading the chapter on the brain (Chap. 4);

To Thomas Karl, for many fruitful discussions and for helping to finalize the figures shown in this book;

To Mark Rinnerthaler, for performing and documenting the experiments shown in Fig. 12.1;

To Elizabeth Rankin and Nicola Santamaria, for their help with the linguistic quality and for critically reading the entire manuscript;

To Rafaela Fürlinger, for her research assistance;

To Tanja Weyandt, for her patient and professional help with the production of the book.

Finally, we want to thank our families for their immeasurable support during the time we spent working on this book project.

Contents

About the Authors

Michael Breitenbach is a retired full professor of genetics at the University of Salzburg in Austria. He has published 230 research and review papers and edited more than 10 books in the fields of the genetics of aging and stress response. He has received a number of national and international honors for his work in genetics, including being elected as a fellow of the AAAS. His main interest in research in the last 30 years has been the utilization of the yeast genetic system to study basic eukaryotic molecular biology, in particular the mechanism and the results of oxidative stress.

Elisabeth Kapferer holds a doctoral degree in German literature studies. She is member of the research staff at the Centre for Ethics and Poverty Research at the University of Salzburg, Austria. Her research activities, which include several (co-)authored and (co-)edited publications, focus on poverty and social exclusion in wealthy societies. Her main research interests are representations of poverty in the arts, public discourse on poverty, and poverty-related disparities in education and health. She is member of the scientific advisory board of the Austrian Health Promotion Fund (Fonds Gesundes Österreich).

Clemens Sedmak holds doctoral degrees in philosophy, theology, and social theory. He is professor of social ethics at the Keough School of Global Affairs and concurrent professor of theology at the University of Notre Dame, IN. He is also co-director of the Centre for Ethics and Poverty Research, University of Salzburg, Austria. He works at the intersection of poverty studies and ethics.

His recent publications include *The Practice of Human Development and Dignity*, co-edited with Paolo Carozza, Notre Dame: University of Notre Dame Press 2020 and *Subsidiarität: Tragendes Prinzip menschlichen Zusammenlebens*, with W. Blum, H. P. Gaisbauer, Regensburg: Friedrich Pustet 2021.

Abbreviations

ACTH	Adrenocorticotropic hormone
Atf4	Activating transcription factor 4
ATP	Adenosine triphosphate
bp	Base pairs
CHD	Coronary heart disease
CRH	Corticotropin-releasing hormone
DMR	Differentially methylated region
EIA	Enzyme immunoassay
eIF2alpha	Eukaryotic initiation factor 2 alpha
ER	Endoplasmic reticulum
GABA	Gamma-aminobutyric acid
GAS	General Adaptation Syndrome
HNE	4-Hydroxynonenal
HPA axis	Hypothalamic–pituitary–adrenal axis
Hsf1	Heat shock transcription factor
HSF1	Yeast essential gene
Hsps	Heat shock proteins
IGF2	Insulin-like growth factor 2
ISR	Integrated stress response
LTD	Long-term depression
LTP	Long-term potentiation
MCP-1	Monocyte chemoattractant protein 1
MIPS	Mental Stress Ischemia Prognosis Study
MMP-9	Matrix metalloproteinase 9
NADH	Reduced nicotinamide adenine dinucleotide
NADPH	Reduced nicotinamide adenine dinucleotide phosphate
NO	Nitric oxide
PNAS	*Proceedings of the National Academy of Sciences*

PNI   Psychoneuroimmunology
PSS   Perceived Stress Scale
PTSD   Post-traumatic stress disorder
RNS   Reactive nitrogen species
RONS   Reactive oxygen and nitrogen species (ROS + RNS)
ROS   Reactive oxygen species
SES   Socioeconomic status
SH   Sulfhydryl
SOD   Superoxide dismutase
SRRS   Social Readjustment Rating Scale
SSRI   Specific serotonin reuptake drugs
TICS   Trier Inventory for the Assessment of Chronic Stress
TSST   Trier Social Stress Test
Yap1   Yeast activating protein 1
YAP1   Yeast gene coding for Yap1

List of Figures

1

Stress and Poverty: An Introduction

"I wish I could press 'pause' at times, I wish I could have a break, relieve the pressure," writes Undine Zimmer (2013, our translation), describing her situati on as a young adult whose background is framed by poverty and poverty-related stress. She continues: "Yet that's not possible. And will not be for the coming years either." Undine Zimmer was born in 1979 and grew up in Berlin, Germany, the daughter of parents who had separated around the time of her birth and were both dependent on welfare payments. Undine was one of so many children, even in affluent societies, who are raised under severely deprived conditions, one of so many people living in poverty. While by the time she wrote down her memories she had managed to overcome most of the restrictions of her childhood, it becomes very clear in her recall that the circumstances in which she found herself as a child and as an adolescent have left their mark—and they still affect her considerably as an adult.

These "circumstances" are not just Undine's individual circumstances. They are shared by people suffering poverty in different countries and societies, even in different times. They would not be sufficiently described solely in statistical terms of economic or material deprivation, such as a low household income, for instance. Rather, these circumstances also involve a particular set of experiences: to point out just a few, frustration, anxiety, fear of what might happen next, struggle and strain, and not least of all, stress.

"Everybody experiences stress, regardless of class," concedes Scottish journalist and musical artist Darren McGarvey (2017), but there is evidence that there are considerable status-related differences in "the degree to which stress inhibits our progress, harms our health and social mobility and shapes our

M. Breitenbach et al., *Stress and Poverty*, https://doi.org/10.1007/978-3-030-77738-8_1

social attitudes and values." In a nutshell, such poverty-linked stress is what this book is about.

Stress is universally experienced. Stress can even be described as not only unavoidable but a vital ingredient of human life. In this sense, stress and relaxation are two poles between which life can develop and thrive, physically, mentally, and emotionally. Poverty thwarts the chance for relaxation. Poverty causes "suffering in body, mind and heart" (Bray et al. 2019). Even times of idleness do not allow for a real pause. William T. Vollmann, in his substantial collection of stories and experiences of "Poor People" (2008), points out that even though people living in poverty and deprivation might live "through hours of nothing to do, their idleness never equal[s] leisure." Among other common dimensions of experiencing poverty, this holds true, no matter where in the world, as a recent international participatory research project led by the international movement ATD Fourth World[1] and the University of Oxford has been able to show (Bray et al. 2019). There is no "pause" key, like the one Undine Zimmer longed for (Zimmer 2013). Or to put it another way: "For those living in poor social conditions … stress is all-consuming; it's the soup everyone is swimming in all the time. Stress is the lens through which all of life is viewed" (McGarvey 2017).

Joining Perspectives on Stress and Poverty

It is our intention in this book to bring together and into conversation different academic perspectives on the linked phenomena of stress and poverty. A deeper understanding of these perspectives is unfurled in the following chapters, which consequently are organized along two lines of reasoning about stress and poverty.[2] Chapters 2–6 are grounded in stress research as developed in a broad array of natural sciences. Chapters 8–12 enter this conversation from the more sociological direction of poverty studies and ethics. Both perspectives contribute important insights to a better understanding of the experience of stress and poverty—and, in particular, of poverty-related stress. These different viewpoints lead to societal and political questions that are at the core of our "cross-disciplinary investigation of stress in cells, individuals, and society" (thus the subtitle of our book). First insights into concepts, definitions, and processes guiding this investigation will be outlined here in the three sections below, *Approaching poverty, Introducing biological stress research,*

---

[1] The acronym "ATD" in the NGO's name ATD Fourth World stands for "All Together in Dignity."

[2] For a more detailed overview of the book, see below for the section, *The Structure of the Book.*

and *Good stress, bad stress, and poverty*. While some of the details presented are complex, they are crucial to the main aim of our book: to advance an interdisciplinary dialogue on the value of the connections between stress and poverty in both academic and policy spheres.

It is necessary for the success of our endeavor, to also investigate and describe the concept of stress as it is used today in the various sciences and in everyday language—this "bridge" in our book is established in Chap. 7 in rich detail. It is an essential question, whether there is a meaningful commonality shared between all those diverse instances where we speak of "stress," ranging from physical/environmental stress, such as **oxidative stress**[3] in single-celled organisms, to chemical stress found in the hormone system and in the neurons of the brains of higher organisms, and finally to the experience of psychological stress, which in our case is the pivotal experience of chronic stress caused by poverty.

This commonality, to anticipate one of our important general findings, is to be found in the biochemical makeup of stressed cells, in the response of those cells to external and internal stresses, and, most importantly, in the regulation of **responses to stress**. This commonality is documented in an overwhelming number of scientific publications, in fields ranging from the genetics and biochemistry of microorganisms to the genetics and biochemistry of the neuroendocrine system of humans and other mammals (in particular the mouse), including psychology, psychiatry, and sociology. As the phenomenon of oxidative stress turns out to be particularly relevant in investigating both physical and psychological stress, first insights into this probable unifying concept are presented here in the introduction (see section "Oxidative Stress as a Unifying Concept" below).

As we will see, there is a growing literature of human stress research that takes into account the idea that there is a relevant interplay between stress and stress responses on the one hand and the **socioeconomic status** of the people experiencing stress on the other. There is also a growing body of literature from multidisciplinary poverty research studying adverse experiences related to poverty, such as stress, and the resulting outcomes for health, especially cardiac and mental health.

---

[3] Glossary terms are bolded at first mention in each chapter.

Approaching Poverty

Poverty is a phenomenon prevalent in almost all countries of the world, a global issue by no means restricted to developing countries. Poverty is prevalent also in developed and highly developed countries, despite enormous technological development and rise in average incomes (Lister 2021; Jefferson 2018). Poverty can be addressed academically from various starting points, from different perspectives, with different assumptions, and with quite heterogeneous objectives in mind. There is a huge body of literature on poverty dating back well into earlier centuries; probably the first broad systematic investigation was initiated by Charles Booth in 1880s London (Spicker 1990; Lepenies 2017). This not only suggests that poverty has been recognized as a "problem" for a long time, but in its diversity also gives evidence that poverty is a phenomenon full of complexity and differing, sometimes even contradictory, aspects and dimensions—as are the lives of human beings.

For the purposes of this book, we will not go into detail about academic discussions. In fact, in this introduction, we will only comment briefly on definitions and measures as well as their societal and political implications. We will instead outline our approach to poverty, which is guided by a genuine interest in the reality of lives lived under conditions of, or at risk of, poverty and on "inner perspectives" of poverty, social marginalization, and **social exclusion**.

As mentioned, there is a long-standing debate about how poverty is best described and measured. Academically, there are different approaches to defining poverty, in terms of absolute or relative poverty, for instance, as well as different ways of measuring poverty, focusing on different tangible or intangible "goods," such as income, assets (and access to assets), and living conditions, but also basic needs, rights, opportunities, and capabilities, to name just a few, and admittedly very diverse, possible starting points. One reason the approaches differ is because poverty occurs in specific contexts, and these contexts may vary depending on time, place, and surroundings; i.e., from biological and geographical conditions as well as from historical, cultural, and societal backgrounds. Regarding definitions and measures, there is as yet no definite and final answer to the question of what poverty is and how it can best be assessed (Walker 2019). Accordingly, this issue continues to be extensively debated, with the aim of addressing and answering these questions more adequately (for a comprehensive overview see Wisor 2012; see also O'Connor 2016; Smeeding 2016; Jefferson 2018; as well as the collections of essays edited by Gaisbauer et al. 2019a or Beck et al. 2020). What we can say, however, is that some definitions and measures may be more context-dependent than others—lack of income, for example, or material deprivation may be the approaches to assessing poor living conditions that are particularly useful in consumption-oriented societies, but these are not so applicable elsewhere. A focus on capabilities (and capability deprivation) on the other hand, i.e., what a person can actually achieve with her available resources (including, yet not restricted to, income), what a person *can do and be* as well as can *choose with good reasons and value to do and be*, offers a far more universal approach (Sen 1980, 1999; Hick and Burchardt 2016).

Nevertheless, poverty always occurs in specific societal contexts. Poverty is not something *unavoidable* by any society but is a *social problem* (Sedmak 2005). As such, poverty is perceived as a situation calling for action and, hence, for informed decisions. Thus, we have to consider that academic discussions of poverty definitions and measures are not just "academic" debates or debates in the often-quoted "ivory tower," but are highly sensitive political issues: for outcomes and conclusions may inform (social) policymakers and have an influence on (social) policy decisions and measures, to name just one area of action (Marlier and Atkinson 2010; Gaisbauer and Sedmak 2014; Gaisbauer and Kapferer 2016; O'Connor 2016; Gaisbauer et al. 2019b; Alkire 2016). As an example, we may think here about agreements on poverty lines that may themselves inform agreements on thresholds of eligibility for social welfare measures. Not least, the issues addressed may also be influential in forming public opinions on "poverty" and "the poor." At the same time, both policy decision-making and public opinion can have an immense impact on the experience of poverty for people actually living in poverty conditions.

The pivotal term here is "experience." Experiencing poverty means experiencing adversity, regardless of the definition or measure applied. Poverty, as we understand it in the context of this book, is an experience forced on people. Poverty in our understanding does not mean the voluntary, deliberate individual choice to abstain from certain goods or activities, but situations and living conditions that induce, in the face of lack of fundamental resources, certain experiences of exclusion and deprivation. Hence, poverty is not merely an abstract issue of interest for research, but an actual daily reality encountered by actual people. The realities of poverty may vary depending on context (and so do definitions and measures, as we have already stated), yet they involve a range of experiences that are shared, regardless of time and place. Not all dimensions of experiencing poverty are obvious; some are "hidden" (Bray et al. 2019), in the sense that they are anxiously concealed, for instance, pointing to the nexus of poverty and shame. Poverty, or social exclusion, "makes it difficult for a person, in Adam Smith's classic words, 'to go about without shame'" (cited in Sedmak 2016; see also Chap. 9). The experience of poverty-related shame, as well as shaming by others, has repeatedly been shown to be "prevalent in both the Global North and South" (Walker 2019; see also Walker 2014). Coping with and making efforts to avoid poverty-related shame and shaming is often also closely linked to the experience of poverty-related stress, as so many firsthand accounts make clear. A focus on such painful experiences seems essential, not least because, as Robert Walker emphasizes, it is "likely to refocus debates away from mere numbers to what really matters, namely people, their experiences and their feelings" (Walker 2019).

Experiences shape identities. Experiences of poverty can have detrimental effects on the development of identities. Poverty not only implies a lack of assets but also a lack of important **life-world** spaces, exclusion from standardized cultural activities, exclusion from access to institutions and systems (including, for instance, education, health care, or the economy) as well as deprivation of capabilities—taken together, a deprivation of **identity resources**. Such resources that allow a person's identity to grow and to flourish include a sense of belonging (being part of an identifiable group); recognition and acknowledgment of one's self received from others; a coherent narration of one's own unique and single life story (with no need to leave things out); and reliable structures of care and concern, together with stable and supportive relationships (Sedmak 2013, 2016). Also, assets and access to assets—"things" as Miller (2008) says—contribute to identity. Poverty disrupts all of the above (Sedmak 2016).

Poverty imposes a challenging and painful experience and a reality that constrains and damages the lives of millions of people worldwide (Lister 2021), raising the question of what kind of poverty knowledge and whose expertise is considered important. As Ruth Lister emphasizes, it is crucial to "acknowledge that, in addition to traditional forms of expertise associated with those who theorize and research poverty, there is a different form of expertise *borne of experience*" (Lister 2021, emphasis added). We will integrate this kind of expertise of poverty knowledge throughout our book in order to learn about poverty as a stressful experience, and as a potential chronic **stressor** (see especially Chap. 9, and subsequent chapters). Insights *from within* and firsthand reports and testimonies are tremendously important sources for our understanding of the experience of poverty.

Certain aspects of poverty-related experiences are part of the life reality we all share to a certain extent as human beings, even if we are not "poor." Such experiences for instance include "**vulnerability**, experience of limited resources [be they material or immaterial, tangible or intangible, we would like to add] and thus restricted choices, the experience of social and cultural pressure, the experience of partial exclusion" (Sedmak 2005). Such experiences, including stress, as Darren McGarvey has indicated above, are not restricted to people of "low socioeconomic status" or to "the poor." Living in poverty, however, may dramatically increase the probability, the load, the intensity, and the persistence of such experiences as well as the potential damage involved. In summary, poverty intensifies the experience of stress.

Introducing Biological Stress Research

To come closer to a general and useful definition of stress we have to consider some basic properties of life. Life as we know it occurs only in living cells. Cells are only alive as long as they maintain an electrochemical potential gradient across the plasma membrane, which envelops the cell. Living cells also maintain an electrochemical potential gradient across the inner mitochondrial membrane and other membranes in the eukaryotic cell. A single hole that allows the unregulated flow of ions across the membrane and is not immediately sealed again is enough to kill a cell. Some antibiotics, such as the gramicidin and amphotericin structural classes, show this beautifully (for review, see Mesa-Arango et al. 2012). Another example is the mode of action of proteins of the **complement system**, which form a **membrane attack complex**, opening a channel or hole in the plasma membrane of an invading pathogen (Heesterbeek et al. 2019). Cell death by programmed suicide (**apoptosis**) likewise uses this principle, although in a different form. A key step in programmed cell death, which is a physiological and, in multicellular organisms, often life-saving process, is the opening of channels in the mitochondrial inner membrane, thereby destroying the electrochemical potential gradient of **mitochondria** and triggering the process of programmed cell death (Lodish et al. 2016).

Concentrations of metabolites inside the cell are not constant but oscillate around a set value, which is maintained by cellular metabolism in a steady state. There is a constant highly regulated flow of substances and metabolites between the cell and its environment by means of transporters and channels, a flow that is at the same time highly specific and efficient and travels in both directions. The purpose of this regulated exchange of ions is to maintain the asymmetric distribution of ions and thereby the electrochemical potential gradient across the cellular membranes.

This situation is generally described as **homeostasis** or **eustasis**. Any large transient deviation (stress, **allostasis**, or **cacostasis**) (Tsigos and Chrousos

2006) from the set value triggers a response to re-establish that value—the stress response. Modern stress research in cell physiology is to a large part occupied with the elucidation of the mechanisms of stress response in answer to the many possible stressors.

Good Stress, Bad Stress, and Poverty

The cellular picture of stress we have described above amounts to a first definition of the concept. More sophisticated definitions of stress will be introduced throughout this book when we review stress research published by numerous authors. However, even at this introductory level, we need to introduce auxiliary concepts and descriptive terms in order to understand the functioning of a live cell under stress. If the external stress applied is mild and transient, it leads to a stress response, which not only causes the return of a parameter (say, concentration of a metabolite) to the set value but also changes the overall pattern of **gene expression** and therefore the **metabolome**, so that a future stressor is tolerated more easily. In such a case, some authors speak of "**eustress**" and more generally of "**hormesis**." The concept of hormesis is generally used and accepted in the physiological literature (Rattan 2008). An easy-to-understand example is the application of a short and mild heat stress (34 °C, 3 h) to yeast cells, resulting in a much-increased resistance to an otherwise lethal heat stress (52 °C, 30 min). Acquiring this resistance depends on the expression of a network of genes, including genes coding for **heat shock proteins (chaperones)** (Hendrick and Hartl 1993). We want to suggest here as a working hypothesis that this concept of hormesis is also at the root of **resilience** to psychological stress, which is nowadays such a pre-eminent topic in psychological stress research (Underwood 2018). However, if the cells are maintained in an unphysiological yet not lethal condition, they may attain a new equilibrium value deviant from the normal physiological one. In this state, the physiology of the whole cell changes, and an increased vulnerability to lethal stressors may result, along with prominent other changes. This may be comparable to chronic psychological stress exemplified by a chronic activation of the neuroendocrine stress system leading to various pathologies for the stressed person, such as "detrimental effects on a variety of physiologic functions, including growth, reproduction, metabolism and immune competence, as well as behavior and personality development" (Tsigos et al. 2016).[4]

---

[4] This "NCBI bookshelf" online review paper on Stress, Endocrine Physiology and Pathophysiology is an Internet publication that is updated at regular intervals. At the time of writing, the most recent update was from 2016.

Such a chronic stress situation is likely to occur under conditions of chronic poverty, thus underlining the significance of our investigation—and we will come back to some of the effects mentioned here in later chapters. Some authors label this kind of stress "**distress**" (Selye 1976). In this sense, poverty can be understood as a probable and multifaceted set of stresses or strains that lastingly exceed the equilibrium value. Overexposure to a (chronic) stressor, e.g., the chronic stress of poverty, can lead to an accumulation of the so-called "**allostatic load**," causing not only mental disturbances but also leading to potentially harmful physiological stress responses (McEwen 2016; McEwen and McEwen 2017).

Indeed, as Sullivan and colleagues state, living in "socially- and physically-disordered" (or "disadvantaged") conditions or neighborhoods that are "characterized by poverty may act as a chronic stressor that results in physiological dysregulation of the body's reactivity to stress" (Sullivan et al. 2019). While research in some cases (e.g., in Sullivan et al. 2019) acknowledges that poverty can come to be a chronic stressor, little attention has been paid to the question of what it actually is that makes poverty a stressor with potential detrimental effects to a person's physical as well as mental health—a "**toxic stressor**," as we call it. In order to understand poverty as a stressor, we need insights into what living in a chronic state of poverty actually means and what this does to a person. This, again, is in particular an inquiry into experiences.

It is important to keep in mind that thinking about experiences is by no means just something that concerns the "individual." Experiences—here, of poverty and related stress—are not least experiences framed and shaped by the respective surroundings and environments, including the respective society a person finds herself in. Experiences of poverty are not to be seen as isolated individual concerns (or worse, faults) but need to be discussed in the context of societal and hence also political issues and questions of opinion- and decision-making; some of them will be addressed in later chapters of this book.

Oxidative Stress as a Unifying Concept

A central role in the stress biochemistry of single-celled as well as higher organisms comes from the phenomenon of oxidative stress, which consequently also plays a role in our investigation into the stress response and human experiences with regard to poverty. Thus, we briefly introduce this phenomenon here, before turning to the structure of the book in the final section of this introduction.

To explain the importance of oxidative stress, we first define it, then explain its origins, and then present evidence that it is very common and nearly universal. A useful definition of oxidative stress that we will draw on again in later chapters is given by Lushchak (2014): "Oxidative stress is a situation when steady state **ROS** (**reactive oxygen species**) concentration is transiently or chronically enhanced, disturbing cellular metabolism and its regulation and damaging cellular constituents."

Oxidative stress is so common because the primordial living cells on earth developed in an anaerobic atmosphere and therefore originally had no defense against oxygen toxicity when oxygen appeared as a product of oxygenic photosynthesis. Oxygen was originally toxic to the cells, but after an ecological catastrophe killing nearly all living cells (Halliwell and Gutteridge 2007), detoxification of oxygen was invented by evolution, i.e., developed by the Darwinian process of mutation and selection, and still later the use of oxygen in respiration became a new source of chemical energy much more efficient than the previously existing forms of glycolysis and fermentation. However, the chemical reactivity of oxygen and of ROS remained and with it the detrimental consequences of the oxidation of proteins, DNA, and RNA. Most extant organisms are nowadays accustomed to the current oxygen concentration in air (about 21%); however, oxygen toxicity still prevails and is easily demonstrated experimentally and through some tragic misconceptions in medical treatments (Halliwell and Gutteridge 2007). Most recently, as evolution continued, ROS, and in particular the non-radical reactive oxygen species, hydrogen peroxide, became a second messenger substance important for regulation of metabolism and cell differentiation, in experimental animals and also in human cells (Sies 2017; Sies et al. 2017). Taken together this means that oxidative stress is an unavoidable fact of aerobic life, and for this reason oxidative stress defense has been conserved through evolution up until today.

In many cases research has discovered that other sources of external stress (for instance, heat) often lead to the parallel development of oxidative stress (Jarolim et al. 2013). More importantly for the purpose of this book, when over-activated, the cells of the central neuroendocrine stress system in humans and other higher organisms create oxidative stress in the neuroendocrine cells of the **hypothalamus** (Tsigos et al. 2016) and in neurons of the **hippocampus** (McEwen 2007). In chronic stress situations, this can lead to neuronal cell death through apoptosis with consequences to be discussed in more detail in chapters to follow. Hence it appears plausible that oxidative stress can be a unifying concept in stress research, including psychological stress in humans caused by the experience of poverty.

As the founding fathers in the field of stress research stated in the 1940s, the main and most important module in the human stress system is the **hypothalamic–pituitary–adrenal axis** (**HPA axis**) (Tsigos and Chrousos 2002). This "axis" links psychological (cognitive) factors with the ancient "fight or flight" response of our early forefathers in situations of immediate life-threatening danger. The HPA axis does not act in isolation, but on the contrary, both the secretion of **cortisol** by the **adrenal cortex** and the secretion of **epinephrine** by the **adrenal medulla** must act together. Epinephrine secretion is triggered by **norepinephrine**, which is not only a hormone but also a neurotransmitter, under the control of the sympathetic nervous system. The interaction of the two hormones, cortisol and epinephrine, leads to the full-blown stress response, consisting among other things, of an increase in heart rate, increase in blood pressure, degradation of the reserve carbohydrate, glycogen, in the muscles with a concomitant increase in serum glucose, an increase in gluconeogenesis, and the metabolic degradation of fat. Figure 1.1 illustrates this combined system.

To sum up, the stress response system served to support the survival of "primitive man" under dangerous conditions. The system is still operating in people today, under very different conditions. In principle, it is still lifesaving but can also have very detrimental consequences under conditions of modern life.[5] These days, activation of the HPA axis and hence oxidative stress can become chronic and thus pathological, causing a variety of diseases that were first described as a common syndrome by Selye (1936, 1955), and also creating severe impediments to a person's development. We will discuss these consequences in the following chapters with a particular focus on the chronic stress of poverty and on the resulting implications in terms of critical social, moral, and political questions.

The Structure of This Book

Let us now turn to the structure of this book. We start our investigation into stress and poverty in Chap. 2, "Hans Selye and the Origins of Stress Research," which offers a brief introduction to the beginnings of the stress research era.

---

[5] This system is not only there for acute stress response, but is necessary for life, as is shown by clinical syndromes lacking cortisol and **aldosterone**, such as congenital adrenal hyperplasia, which lead to death in early childhood (Speiser and White 2003).

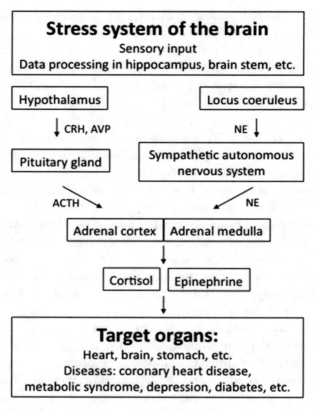

Fig. 1.1 Introduction to the stress system. The two branches of the stress system sketched in this figure work together to produce a fully developed stress response. See text for the differential action of cortisol and epinephrine in the stress response. The figure is a simplified scheme of the stress response as, for example, the parasympathetic nervous system, and the detailed regions of the brain which are involved have been omitted. A more detailed description of our present knowledge of the brain centers active in the stress response is given in Chap. 4. CRH, corticotropin-releasing hormone; AVP, arginine vasopressin; NE, norepinephrine; ACTH, adrenocorticotropic hormone

Admittedly, there was groundbreaking work on traumatic shock and emotional excitement before Selye, early in the twentieth century, with explorations of physiological as well as psychological processes involved in the experience of strong bodily and mental stimuli, e.g., in the pioneering works of Walter Bradford Cannon (Cannon 1915). However, it was not until the work of Hans Selye that the nowadays widely used term "stress" emerged and developed into a multifaceted concept that included both threatening an organism through "**allostatic overload**" (McEwen 2005) and providing for an organism's positive and healthy development through the concept of "hormesis."

After this brief historical section, in Chap. 3, "Oxidative and Other Stress Research at the Cellular Level," we offer a detailed introduction to "oxidative stress," a description of central reactions in cells provoked by oxidative stress, and central features of the response to oxidative and related types of stress. This chapter offers a deep dive into biochemical stress research: it provides insights into what in evolution is arguably the oldest stress that organisms are exposed to; it also looks into defense pathways (leading to the evolution of respiration, for example) as well as various signaling pathways. We see that oxidative stress and the oxidative stress response are core factors that allowed for the evolution of life on earth up to the present day. Again, we see that stress is an ambivalent phenomenon. While producing detrimental effects on organisms, stress is also essential for the survival of organisms, from single cells to human beings.

Chapter 4, "Oxidative Stress and the Brain," discusses the outcomes of oxidative stress on the brain. Here we shift our focus to the destructive effects of chronic stress on different regions of the brain, describing the manifestations of these effects in a number of stress-related syndromes and diseases, including **major depression**, psychosomatic heart diseases, and the impact of stress on the immune system. Research into such adverse outcomes of chronic stress exposure clearly touches the psychological side of stress, which will be a crucial aspect later, when we enter into a conversation between stress research and poverty studies.

Chapter 5, "Epigenetics and Some Further Observations on Stress-Induced Diseases," discusses the role of epigenetics in stress research, focusing on a comparably recent strand of research. We present some well-studied and striking examples of epigenetic processes caused by stress, such as hunger stress, caregiving stress, and stress experienced during pregnancy. This section of the book also discusses certain types of heart disease linked to psychological stress, where epigenetics comes into play.

After having highlighted selected aspects of stress research that show its richness and depth, we address a major issue in the field: the crucial question of how to measure stress. Numerous methods have been developed in order to quantify stress. Chapter 6, "Measuring Stress," offers a tour through the wide landscape of approaches targeting acute and chronic stress, considering minor daily stressors as well as major life events, and using analytical, biochemical, and physiological methods on the one hand and psychological methods on the other. We also outline how stress research developed, expanding from experiments *in the laboratory* (in cells, such as yeast cells, and in animals, such as rodents, as well as in humans) to setups *in the field*, which aim at reliably assessing stress levels under conditions of real, everyday life—including often chronic real-life conditions shaped by poverty.

Chapter 7, "The Language Games of Stress," invites the reader to pause for a moment and reflect on the way the term "stress" is used. This reflection seems necessary since the book is the result of an interdisciplinary collaboration between biological stress research and poverty research. Furthermore, "stress" is a term that is used in ordinary language as well as in academic discourse. We argue that "stress" is used in these different contexts in ways that show similarities or "family resemblances." The concepts of oxidative stress and of psychological stress show sufficient common ground to justify the meaningful use of the term "stress" across different contexts and disciplines.

The subsequent chapters are dedicated to exploring the link between stress and poverty in more detail. We attempt to show how poverty can very likely result in chronic and "toxic" stress, i.e., in stress a person is unable to avoid, to resolve, or to overcome. We begin this part by first reconstructing how poverty became recognized as a relevant topic for stress research. Chapter 8, "The Unhealthy Relationship Between Stress and Poverty," shows how poverty and poverty-related stress lead to more adversity and more stress, forming a vicious circle. While corroborating findings from previous chapters, this chapter also demonstrates that the relationship between stress and poverty is unhealthy because of the numerous adverse health outcomes that have been identified by stress research.

Poverty, as we argued earlier in this introduction, refers to the experience of adverse, involuntary, and stress-filled living conditions. With poverty-related stress, it is difficult to see any positives. In Chap. 9, "The Stressful Experience of Poverty," we emphasize the adverse conditions of poverty by considering firsthand accounts of the experience of poverty. We learn from these personal testimonies how poverty and poverty-related stress shape people's minds, their options, choices, opportunities, and prospects for the future.

Beyond this personal side of the experience of poverty, we can observe social, moral, and political aspects of poverty and poverty-related stress. These dimensions are taken up in the remaining chapters.

A sociological reading of stress is provided in Chap. 10, "Social and Moral Aspects of Stress." The experience of stress takes place in social settings, in certain surroundings, within certain political frameworks. The environment of the experience of stress is shaped by power dynamics, including social and cultural norms, and dynamics of vulnerability. Poverty, as a source of increased vulnerability levels, is very likely to deepen power imbalances, which in turn lead to higher and more consolidated stress levels.

Chapter 11, "The Politics of Stress," discusses the political dimensions of stress and poverty. We take politics to be the art of managing power, and in particular of managing the power to shape environments and frameworks and to impose normative expectations. We pay special attention to the idea of social order, to the question of stress distribution, and to the role of institutions. This brings us to an important question: how can we build *stress-sensitive* institutions and *stress-sensitive* societies—that is, institutions and societies that in structural terms minimize the risk of toxic stress exposure? Chronic stress, such as poverty-related stress, results in a harmful disruption to the organism and to an environment and order that allows for functioning and flourishing, as can be seen in the biochemical model as well as in the sociological model of stress.

The question of how to build stress-sensitive environments leads us to the question of response and coping mechanisms and thus to the question of resilience—the ability to return to a state of functioning, the capacity to maintain stable functioning in the face of adversity, and the ability to adapt to new conditions. "Resilience," just like stress, proves to be a fruitful concept for our project, in that resilience bridges discourses of biochemistry, social sciences, and social ethics. Chapter 12, "Responding to Stress and the Value of Resilience," argues that resilience research and poverty research can enter a fruitful dialogue by exploring the pressure on poor people to be resilient even while they are living in poverty conditions that undermine sources of resilience. Just as in the case of stress, there is the moral and political issue of unequal distribution—of access to resources of resilience and of the pressure to be resilient. These observations lead to building blocks for a "politics of resilience" and to pathways toward stress-sensitive societies.

While we were writing this book, the world was turned upside down in 2020 by the pandemic caused by the coronavirus SARS-CoV-2 (or COVID-19). This crisis caused unprecedented stressful conditions and high stress levels. In many regards, the phenomena that we have described in our book have been brought into focus by the pandemic as if seen through a magnifying glass. We end our book with Chap. 13, with an "Epilogue: The Pandemic as a Big Reveal," where we discuss some of the key concerns of this book related to the connection between stress and poverty, in the light (or darkness) of the pandemic. There can be no doubt that a global crisis like that of COVID-19 reveals a great deal about the tight links between stress and poverty.

The main motive for this joint, multidisciplinary investigation into stress and poverty and the intersections between the two is our concrete shared interest in chronic poverty as a critical, toxic stressor. Writing this book together, we have aimed at enabling a fruitful conversation across disciplines—fruitful in the sense of shedding light on the dark experience of chronic and toxic poverty-related stress and its detrimental consequences at the individual and societal levels. Our interest is by no means solely theoretical or academic. In fact, this topic leads directly to the realm of policy change and to the possibility of research-informed interventions that are sensitive to issues of both poverty and stress. We hope that we can contribute to both a new awareness of the stressful experience of poverty as well as to pathways that hold promise for mitigating this experience for people living in poverty.

References

Alkire A (2016) The capability approach and well-being measurement for public policy. In: Adler MD, Fleurbaey M (eds) The Oxford handbook of well-being and public policy. Oxford University Press, Oxford. https://doi.org/10.1093/oxfordhb/9780199325818.013.18

Beck V, Hahn H, Lepenies R (eds) (2020) Dimensions of poverty. measurement, epistemic injustices, activism. Springer, Cham

Bray R, De Laat M, Godinot X, Ugarte A, Walker R (2019) The hidden dimensions of poverty. International participatory research. Fourth World, Montreuil

Cannon WB (1915) Bodily changes in pain, hunger, fear and rage: an account of recent researches into the function of emotional excitement. D. Appleton, New York

Gaisbauer HP, Kapferer E (2016) Suffering within, suffering without: paradoxes of *poverties* in welfare states. In: Gaisbauer HP, Schweiger G, Sedmak C (eds) Ethical issues in poverty alleviation. Springer, Cham, pp 171–189. https://doi.org/10.1007/978-3-319-41430-0_10

Gaisbauer HP, Sedmak C (2014) Neglected futures. Considering overlooked poverty in Europe. Eur J Future Res 57(2):1–8. https://doi.org/10.1007/s40309-014-0057-2

Gaisbauer HP, Schweiger G, Sedmak C (eds) (2019a) Absolute poverty in Europe. Interdisciplinary perspectives on a hidden phenomenon. Policy Press, Bristol

Gaisbauer HP, Schweiger G, Sedmak C (2019b) Absolute poverty in Europe: introduction. In: Gaisbauer HP, Schweiger G, Sedmak C (eds) Absolute poverty in Europe. Interdisciplinary perspectives on a hidden phenomenon. Policy, Bristol, pp 1–14

Halliwell B, Gutteridge JMC (2007) Free radicals in biology and medicine. Oxford University Press, Oxford

Heesterbeek DAC, Bardoel BW, Parsons ES, Bennett I, Ruyken M, Doorduijn DJ, Gorham RD Jr, Berends ETM, Pyne ALB, Hoogenboom BW, Rooijakkers SHM (2019) Bacterial killing by complement requires membrane attack complex formation via surface-bound C5 convertases. EMBO J 38(e99852):1–17. https://doi.org/10.15252/embj.201899852

Hendrick JP, Hartl F-U (1993) Molecular chaperone functions of heat-shock proteins. Annu Rev Biochem 62:349–384

Hick R, Burchardt T (2016) Capability deprivation. In: Brady D, Burton LM (eds) The Oxford handbook of the social science of poverty. Oxford University Press, Oxford. https://doi.org/10.1093/oxfordhb/9780199914050.013.5

Jarolim S, Ayer A, Pillay B, Gee AC, Phrakaysone A, Perrone GG, Breitenbach M, Dawes IW (2013) Saccharomyces cerevisiae genes involved in survival of heat shock. G3 3:2321–2333. https://doi.org/10.1534/g3.113.007971

Jefferson PN (2018) Poverty. A very short introduction. Oxford University Press, Oxford

Lepenies P (2017) Armut. Ursachen, Formen, Auswege. C. H. Beck, Munich

Lister R (2021) Poverty (key concepts), 2nd edn. Polity, Cambridge

Lodish H, Berk A, Kaiser CA, Krieger M, Bretscher A, Ploegh H, Amon A, Martin KC (2016) Molecular cell biology. W. H. Freeman, New York

Lushchak VI (2014) Free radicals, reactive oxygen species, oxidative stress and its classification. Chemico-Biol Interact 224:164–175. https://doi.org/10.1016/j.cbi.2014.10.016

Marlier E, Atkinson AB (2010) Indicators of poverty and social exclusion in a global context. J Policy Anal Manag 29(2):285–304. https://doi.org/10.1002/pam.20492

McEwen BS (2005) Stressed or stressed out: what is the difference? J Psychiatry Neurosci 30(5):315–318

McEwen BS (2007) Physiology and neurobiology of stress and adaptation: central role of the brain. Physiol Rev 87:387–904. https://doi.org/10.1152/physrev.00041.2006

McEwen BS (2016) Central role of the brain in stress and adaptation: allostasis, biological embedding, and cumulative change. In Fink G (ed) Stress: concepts, cognition, emotions, and behavior. Handbook of stress, vol 1. Academic, Amsterdam, pp 39–55. https://doi.org/10.1016/B978-0-12-800951-2.00005-4

McEwen CA, McEwen BS (2017) Social structure, adversity, toxic stress, and intergenerational poverty: an early childhood model. Annu Rev Sociol 43:29.1–29.28. https://doi.org/10.1146/annurev-soc-060116-053252

McGarvey D (2017) Poverty safari. Understanding the anger of Britain's underclass. Picador, London

Mesa-Arango A, Scorzoni L, Zaragoza O (2012) It only takes one to do many jobs: amphotericin B as antifungal and immunomodulatory drug. Front Microbiol 3:1–10. https://doi.org/10.3389/fmicb.2012.00286

Miller D (2008) The comfort of things. Polity, Cambridge

O'Connor A (2016) Poverty knowledge and the history of poverty research. In: Brady D, Burton LM (eds) The Oxford handbook of the social science of poverty. Oxford University Press, Oxford. https://doi.org/10.1093/oxfordhb/9780199914050.013.9

Rattan SIS (2008) Hormesis in aging. Ageing Res Rev 7:63–78

Sedmak C (2005) Introduction: commitments and an "option for the poor". In: Holztrattner M, Sedmak C (eds) Humanities and option for the poor. LIT, Vienna, pp 9–21

Sedmak C (2013) Armutsbekämpfung. Eine Grundlegung. Böhlau, Vienna

Sedmak C (2016) A church of the poor. Pope Francis and the transformation of orthodoxy. Orbis, Maryknoll, NY

Selye H (1936) A syndrome produced by diverse nocuous agents. Nature 138:32

Selye H (1955) Stress and disease. Science 122(3171):625–631

Selye H (1976) Forty years of stress research: principal remaining problems and misconceptions. CMA J 115:53–56

Sen A (1980) Equality of what? In: McMurrin SM (ed) The Tanner lectures on human values, vol 1. Cambridge University Press, Cambridge, pp 195–220

Sen A (1999) Development as freedom. Oxford University Press, Oxford

Sies H (2017) Hydrogen peroxide as a central redox signaling molecule in physiological oxidative stress: oxidative eustress. Redox Biol 11:1–7. https://doi.org/10.1016/j.redox.2016.12.035

Sies H, Berndt C, Jones DP (2017) Oxidative stress. Annu Rev Biochem 86:715–748. https://doi.org/10.1146/annurev-biochem-061516-045037

Smeeding TM (2016) Poverty measurement. In: Brady D, Burton LM (eds) The Oxford handbook of the social science of poverty. Oxford University Press, Oxford. https://doi.org/10.1093/oxfordhb/9780199914050.013.3

Speiser PW, White PC (2003) Congenital adrenal hyperplasia. N Engl J Med 349(8):776–788

Spicker P (1990) Charles Booth: the examination of poverty. Soc Policy Adm 24(1):21–38. https://doi.org/10.1111/j.1467-9515.1990.tb00322.x

Sullivan S, Kelli HM, Hammadah M, Topel M, Wilmot K et al (2019) Neighborhood poverty and hemodynamic, neuroendocrine, and immune response to acute stress among patients with coronary artery disease. Psychoneuroendocrinology 100:145–155. https://doi.org/10.1016/j.psyneuen.2018.09.040

Tsigos C, Chrousos GP (2002) Hypothalamic–pituitary–adrenal axis, neuroendocrine factors and stress. J Psychosom Res 53:865–871

Tsigos C, Chrousos GP (2006) Stress, obesity, and the metabolic syndrome: soul and metabolism. Ann N Y Acad Sci 1083:xi–xiii. https://doi.org/10.1196/annals.1367.025

Tsigos C, Kyrou I, Kassi E, Chrousos GP (2016) Stress, endocrine physiology and pathophysiology [updated 2016 Mar 10]. In: Feingold KR, Anawalt B, Boyce A et al (eds) Endotext [Internet]. MDText.com, South Dartmouth, MA. Available from: https://www.ncbi.nlm.nih.gov/books/NBK278995/

Underwood E (2018) Lessons in resilience: in war zones and refugee camps, researchers are putting resilience interventions to the test. Science 359(6379):976–679. https://doi.org/10.1126/science.359.6379.976

Vollmann WT (2008) Poor people. Harper Perennial, New York

Walker R (2014) The shame of poverty. Oxford University Press, Oxford

Walker R (2019) Measuring absolute poverty: shame is all you need. In: Gaisbauer HP, Schweiger G, Sedmak C (eds) Absolute poverty in Europe. Interdisciplinary perspectives on a hidden phenomenon. Policy, Bristol, pp 97–118

Wisor S (2012) Measuring global poverty. Toward a pro-poor approach. Palgrave, New York. https://doi.org/10.1057/9780230357471

Zimmer U (2013) Nicht von schlechten Eltern. Meine Hartz-IV-Familie. S. Fischer, Frankfurt a. M.

2

Hans Selye and the Origins of Stress Research

Stress research started with a seminal paper in *Nature* (Selye 1936). Although the concept existed in medicine before that time, it was not called "stress," and Hans Selye was the first to give it a firm footing as a basic endocrinological and medical concept. His idea, based on the case histories of his patients and later on animal experiments, was that a rather large variety of cases could lead to very similar ailments, among them ulcers, heart disease, thymic regression, adrenal hyperplasia, and depression, all accompanied by activation of the hormonal **hypothalamic–pituitary–adrenal axis (HPA axis)**[1] (Tsigos and Chrousos 2002). In this chapter, we offer a brief portrayal of Hans Selye and an introduction to the groundbreaking early developments of stress research that he sparked, thereby strongly inspiring subsequent generations of researchers.

Biographical Note

Hans Selye, or Selye János in his native Hungarian, was born in Vienna in 1907 and spent his childhood days in Komárom (now Komárno in Slovakia). His father was a surgeon colonel in the Austro-Hungarian army. Selye studied medicine at the Charles University of Prague, obtaining his MD in 1929 as well as a PhD in organic chemistry. He moved to Johns Hopkins University in Baltimore on a Rockefeller Foundation fellowship and later moved to the Department of Biochemistry of McGill University in Montreal, where he remained for the rest of his extremely active life of research in physiology and endocrinology until his death in 1982 (Fig. 2.1).

---

[1] Glossary terms are bolded at first mention in each chapter.

© Springer Nature Switzerland AG 2021
M. Breitenbach et al., *Stress and Poverty*, https://doi.org/10.1007/978-3-030-77738-8_2

Fig. 2.1 Hans Selye during his time as a professor of medicine in Montreal (photo taken by Jean-Paul Rioux). Source: Wikimedia Commons (CC BY-SA 4.0). https://commons.wikimedia.org/wiki/File:Portrait_Hans_Selye.jpg

The Beginning of Stress Research in Endocrinology

It was in Montreal that Hans Selye discovered, during his studies with laboratory animals, a syndrome that reminded him of what he had seen in chronically ill patients and which he called "**General Adaptation Syndrome**" (**GAS**). This became the starting point in 1935 to stress research worldwide and led to over 1500 research articles and 35 books authored by Selye himself and more than 100,000 further publications on stress. Selye was obviously a gifted and inspiring writer and spoke ten languages fluently, in addition to his native Hungarian.

In his first paper (Selye 1936), the term "stress" was not even mentioned, although the author did describe the GAS. A primary stress, a major injury like a burn wound, or an infection, for instance, would trigger a response of the psycho-neuro-endocrinological system (in modern wording) which could, in extreme cases, lead to the death of the patient or of the experimental animal. A distinct but overlapping clinical syndrome is the "**cytokine storm syndrome**" (Behrens and Koretzky 2017; Chousterman et al. 2017), leading to death in an unacceptably large percentage (about 20%) of sepsis patients. The complexity of the topic is shown by the fact that even now, after so many decades of research, an effective and safe treatment for sepsis patients is still lacking (Chousterman et al. 2017). In less extreme cases the prolonged **stress response** could induce the formation of gastric ulcers or other severe somatic diseases in the patient long after the wound was healed. What impressed Hans Selye most was: (1) that the response was detached in time and location from the primary stress; and (2) that very different primary stresses could cause very similar final outcomes of the stress response.

During the next 20 years, it was established that the stress response could be studied in mice and rats and many other animals, and basic facts about the biochemical changes triggered by the stress response were elucidated. These responses were to a large degree similar in all higher animals studied (Selye 1955). For instance, the **adrenal hormones**, which are the proximate effectors of the stress response, were the same in rodents and humans, with one exception: while **cortisol** is the main effective **glucocorticoid** in humans, its place in rodents is taken rather by its oxidation product, **cortisone**. However, all the other adrenal hormones are the same in structure and efficacy between mammalian species. The impact of this new science of stress was so great that medical researchers all over the world joined and worked in this field (Selye 1976b). To the present day, medical and sociological stress research is a growing field. Specialized journals and societies were founded, and over a hundred thousand scientific papers on stress physiology and stress as a problem of society have been published to date.

Heat Shock

In addition to the development of stress research in endocrinology and medicine, a second and equally important discovery took place at roughly the same time and eventually led to avenues of research that enabled genetic and later molecular biological access to the biology of stress. This was the discovery of the heat shock response in *Drosophila* (Ritossa 1962; Ritossa 1996). It was shown subsequently that a non-lethal but strong temperature shift (heat shock) induced a change in **gene expression** not only in *Drosophila* but in every type of living cell that was investigated. The heat shock response was found to be very highly conserved in all domains of life and to be activated not only by raised temperature but also by other stresses, even in those organisms that keep the body temperature constant or nearly constant. The original discovery was made before it was even known that messenger RNA (**mRNA**) exists and that **ribosomes** are the machines of protein biosynthesis. However, soon after Ritossa's discovery, molecular biology became an all-pervading paradigm of biological research, and the toolbox of molecular biology was developed step by step and applied to stress (and heat shock) research. We now know that the **chromosome puffs** discovered by Ritossa were gene-specific accumulations of mRNA, indicating that the stress situation induced the expression of new genes and led to adaptation to the new environmental conditions (a higher but still not lethal temperature). The central genes induced in this experiment are the heat shock transcription factors and the additional genes expressed are the target genes of these transcription factors; for instance, genes coding for **heat shock proteins** (**hsps**). The primary function of heat shock proteins is to help proteins to attain their correct three-dimensional folding, which may be lost under stress conditions. Unfolded and aggregated proteins are a major cause of many diseases, especially neurodegenerative diseases (Lodish et al. 2016, in their third chapter). In fact, the genetic program discovered through heat shock research highly overlapped with the genetic program of stress responses in mammals discovered much later (Åkerfelt et al. 2010).

Stress, Hormesis, Aging, Shock

Right from the beginning of stress research, Selye and the authors following him distinguished "**eustress**" from "**distress**" (Selye 1976a). As we have hinted in the introduction to this book, the rational explanation of eustress in modern terms is to be found in "**hormesis**," leading to a gene expression program enabling the cell to cope with a later and larger stress challenge. On the other hand, "distress" is defined simply as a repeated or chronic stress that in the end leads to **exhaustion** (Selye's term) and to severe untoward physiological changes. Selye noticed that the state of exhaustion from repeated or chronic stress (in mice, for instance) "resembled senility" (Selye's words). This is in perfect agreement with modern findings showing that the process of aging is linked to cellular and organismic **oxidative stress** (Aung-Htut et al. 2012, 2013) and invariably leads to a loss of stress resistance (Liguori et al. 2018). Stress resistance is the preferred term in cellular research, closely related but not identical to **resilience**, the term preferred by psychologists and neurologists (Underwood 2018; see Chap. 12).

One of the findings described in the second seminal paper discussed here (Selye 1955) is the in part synergistic and in part antagonistic action of cortisol (glucocorticoid) and **aldosterone (mineralocorticoid)**, although production of both is governed by **adrenocorticotropic hormone (ACTH)**, and both are made in and secreted from the **adrenal cortex**. To give examples, in the regulation of inflammation, the two hormones are clearly antagonistic, while the production of severe kidney damage by a high dose of deoxycorticosterone (a mineralocorticoid interfering with normal sodium excretion) is not counteracted but aggravated by cortisol.

Selye recognized that acute traumatic shock is an example of a general alarm reaction. Shock can have several quite different origins, one being massive infection with certain bacteria; another could be blood loss or another traumatic experience, even a massive psychological shock. Acute shock situations still present a major risk of death for the patient. The immediate cause of death frequently is not the "obvious" one (for instance, overgrowth of bacteria in septic shock) but rather lies in the body's answer to the infection, the so-called cytokine storm (Dare et al. 2009; Shirota et al. 2005; Chousterman et al. 2017). **Cytokines** were not known in 1955 but viewed from today's perspective we know that these peptide hormones are an integral part of the hormone system playing a central role in the inflammation reaction and interacting with the HPA axis in a stress situation. Even today there is no sufficiently safe and secure method to rescue a patient in toxic shock before multiple organ failure occurs as a consequence of a cytokine storm. Selye was primarily a medical researcher trying to understand the causes of diseases. The diseases we are discussing here might in modern language be called stress-induced diseases, but in Selye's early papers they were called diseases of adaptation caused by "errors in the adaptation syndrome."

The measured increase in serum cortisol (and concomitant endocrine changes) in acute or chronic stress leads to an increase in the susceptibility to infection because of the immune suppressive effects of the hormone, which are only today better understood (Straub 2014). This fact can be shown by systematic experiments with mice or guinea pigs, but it is also seen in patients who are treated with corticoids to relieve chronic inflammatory disease, in rheumatoid arthritis, for instance (Straub 2014).

Lifestyle Factors

Modern medical statistical investigations show the enormous influence of lifestyle factors on health span and lifespan, and many of the lifestyle factors monitored are directly influenced by stress physiology (Li et al. 2018). This can be seen from work with people who are under stress in their jobs (Li et al. 2019) or by their situation as single mothers with chronically sick children (Epel et al. 2004, 2006, 2008; Wolkowitz et al. 2011). A clear correlation is found between chronic stress, cardiovascular disease, and depression (Wolkowitz et al. 2011). We also see from the previous literature on poverty and disease (e.g., the World Health Report 2002) that this cluster of symptoms or disease markers is correlated with poverty.

In Selye's words, "stress can both cause and be caused by mental reactions" (Selye 1955). This close connection between psychological stress and the HPA axis was noticed by Selye as he carefully observed patients. From a modern perspective, we can say that these initial observations, dating back to the 1950s, have been completely confirmed. For instance, it was shown that the endocrine status of people suffering from severe depression is very similar to people under acute psychological or other stresses (Yang et al. 2015). It is plausible from studies in human genetics and in neurology that **major depression** is primarily caused by environmental and life-history factors and less so by genetic predisposition (Border et al. 2019), especially when compared with other major psychiatric diseases such as bipolar disease and schizophrenia (Kandel 2018). It is common knowledge that severe psychological stress can cause a range of somatic diseases, perhaps most importantly chronic and acute heart diseases like heart infarction (Lagraauw et al. 2015).

Many of the just-mentioned basic facts of stress physiology are now common knowledge and are used in scientific and popular books and articles without special reference to the author who originated these concepts, Hans Selye.

The work of Hans Selye gained unprecedented international importance and popularity, and his books were translated into all major languages. He not only started a new and important scientific field, but was also a gifted popularizer of science, and the stress research that he started had an enormous influence on the social sciences. In some of his books, Selye dealt intensively with questions of practical ethics and in particular with the question of how in everyday life a person could deal with social stress of different kinds and how one could convert the distress of life into eustress, gaining in maturity and patience and the gift of enjoying life. This process, if it is successful, is equivalent to gaining resilience, as beautifully described in Selye's autobiographical book *The Stress of Life* (Selye 1976a).

References

Åkerfelt M, Morimoto R, Sistonen L (2010) Heat shock factors: integrators of cell stress, development and lifespan. Mol Cell Biol 11:545–555

Aung-Htut MT, Ayer A, Breitenbach M, Dawes IW (2012) Oxidative stress and ageing. In: Breitenbach M, Jazwinski SM, Laun P (eds) Aging research in yeast. Springer, Dordrecht, pp 13–54. https://doi.org/10.1007/978-94-007-2561-4_2

Aung-Htut MT, Lam YT, Lim Y-L, Rinnerthaler M, Gelling CL, Yang H, Breitenbach M, Dawes IW (2013) Maintenance of mitochondrial morphology by autophagy and its role in high glucose effects on chronological lifespan of Saccharomyces cerevisiae. Oxid Med Cell Longev 2013:1–13. https://doi.org/10.1155/2013/636287

Behrens EM, Koretzky GA (2017) Cytokine storm syndrome: looking toward the precision medicine era. Arthritis Rheumatol 69(6):1135–1143. https://doi.org/10.1002/art.40071

Border R, Johnson EC, Evans LM, Smolen A, Berley N, Sullivan PF, Keller MC (2019) No support for historical candidate gene or candidate gene-by-interaction hypotheses for major depression across multiple large samples. Am J Psychiatry 176(5):376–387. https://doi.org/0.1176/appi.ajp.2018.18070881

Chousterman BG, Swirski FK, Weber GF (2017) Cytokine storm and sepsis disease pathogenesis. Semin Immunopathol 39:517–528. https://doi.org/10.1007/s00281-017-0639-8

Dare AJ, Phillips ARJ, Hickey AJR, Mittal A, Loveday B, Thompson N, Windsor JA (2009) A systematic review of experimental treatments for mitochondrial dysfunction in sepsis and multiple organ dysfunction syndrome. Free Radic Biol Med 47:1517–1525. https://doi.org/10.1016/j.freeradbiomed.2009.08.019

Epel ES, Blackburn EH, Lin J, Dhabhar FS, Adler NE, Morrow JD, Cawthon RM (2004) Accelerated telomere shortening in response to life stress. PNAS 101(49):17312–17315. https://doi.org/10.1073#pnas.0407162101

Epel ES, Lin J, Wilhelm FH, Wolkowitz OM, Cawthon R, Adler NE, Dolbier C, Mendes WB, Blackburn EH (2006) Cell aging in relation to stress arousal and cardiovascular disease risk factors. Psychoneuroendocrinology 31:277–287. https://doi.org/10.1016/j.psyneuen.2005.08.011

Epel ES, Stein Merkin S, Cawthon R, Blackburn EH, Adler NE, Pletcher MJ, Seeman TE (2008) The rate of leukocyte telomere shortening predicts mortality from cardiovascular disease in elderly men. Aging 1(1):81–88

Kandel ER (2018) The disordered mind. Robinson, London

Lagraauw HM, Kuiper J, Bot I (2015) Acute and chronic psychological stress as risk factors for cardiovascular disease: insights gained from epidemiological, clinical and experimental studies. Brain Behav Immun 50:18–30. https://doi.org/10.1016/j.bbi.2015.08.007

Li Y, Pan A, Wang DD, Liu X, Dhana K, Franco OH, Kaptoge S, Di Angelantonio E, Stampfer M, Willett WC, Hu FB (2018) Impact of healthy lifestyle factors on life expectancies in the US population. Circulation 137:1–16. https://doi.org/10.1161/CIRCULATIONAHA.117.032047

Li J, Atasoy S, Fang X, Angerer P, Ladwig K-H (2019) Combined effect of work stress and impaired sleep on coronary and cardiovascular mortality in hypertensive workers: the MONICA/KORA cohort study. Eur J Prev Cardiol 1–8. https://doi.org/10.1177/2047487319839183

Liguori I, Russo G, Curcio F, Bulli G, Aran L, Della-Morte D, Gargiulo G, Testa G, Cacciatore F, Bonaduce D, Abete P (2018) Oxidative stress, aging, and diseases. Clin Interv Aging 13:757–772. https://doi.org/10.2147/CIA.S158513

Lodish H, Berk A, Kaiser CA, Krieger M, Bretscher A, Ploegh H, Amon A, Martin KC (2016) Molecular cell biology. W. H. Freeman, New York

Ritossa F (1962) A new puffing pattern induced by temperature shock and DNP in Drosophila. Experientia 18:571–573

Ritossa F (1996) Discovery of the heat shock response. Cell Stress Chaperones 1(2):97–98

Selye H (1936) A syndrome produced by diverse nocuous agents. Nature 138:32

Selye H (1955) Stress and disease. Science 122(3171):625–631

Selye H (1976a) The stress of life. McGraw-Hill, New York

Selye H (1976b) Forty years of stress research: principal remaining problems and misconceptions. Can Med Assoc J 115(1):53–56

Shirota H, Gursel I, Gursel M, Klinman DM (2005) Suppressive oligodeoxynucleotides protect mice from lethal endotoxic shock. J Immunol 174:4579–4583. https://doi.org/10.4049/jimmunol.174.8.4579

Straub RH (2014) Interaction of the endocrine system with inflammation: a function of energy and volume regulation. Arthritis Res Ther 16(203):1–15

Tsigos C, Chrousos GP (2002) Hypothalamic–pituitary–adrenal axis, neuroendocrine factors and stress. J Psychosom Res 53:865–871

Underwood E (2018) Lessons in resilience: in war zones and refugee camps, researchers are putting resilience interventions to the test. Science 359(6379):976–679. https://doi.org/10.1126/science.359.6379.976

Wolkowitz OM, Mellon SH, Epel ES, Lin J, Dhabhar FS, Su Y, Reus VI, Rosser R, Burke HM, Kupferman E, Compagnone M, Nelson JC, Blackburn EH (2011) Leukocyte telomere length in major depression: correlations with chronicity, inflammation and oxidative stress—preliminary findings. PLoS One 6(3):1–10. https://doi.org/10.1371/journal.pone.0017837

World Health Report (2002) Reducing risks, promoting healthy life. World Health Organization, Geneva

Yang L, Zhao Y, Wang Y, Liu L, Zhang X, Li B, Cui R (2015) The effects of psychological stress on depression. Curr Neuropharmacol 13:494–504

3

Oxidative and Other Stress Research at the Cellular Level

In this chapter, we describe the biochemical reactions in **oxidative stress**[1] and oxidative **stress response** as they have been discovered through research with single-cell systems. The reason for including this in our book is twofold: First, oxidative stress is arguably the oldest stress to which living cells are exposed and has been highly conserved in evolution. Second, so are the defense pathways that Mother Nature invented (developed by the Darwinian process of mutation and selection) during evolution. We also describe hypotheses about the origin of oxygen and of respiration and the importance of the metabolic pathways involved in stress generation and stress response during the course of evolution on this earth. This includes the development of signaling pathways based on oxygen radicals and **reactive oxygen species** (**ROS**), which play a major role in the homeodynamic regulation of extant organisms. Moreover, the stress and stress response pathways described here do exist in neuronal cells and are essential in both forming memory and establishing a stressed state in individuals.

Note: For those of our readers who would rather not study biochemical reaction sequences, it is safe to skip this particular chapter and to continue with the subsequent chapters.

---

[1] Glossary terms are bolded at first mention in each chapter.

© Springer Nature Switzerland AG 2021
M. Breitenbach et al., *Stress and Poverty*, https://doi.org/10.1007/978-3-030-77738-8_3

Introduction

The rationale for this part of our book on stress is that we want to understand psychological stress in cells of the central nervous system at the molecular level. From all that we currently know about stress at the cellular level, we see a continuity of mechanisms and phenomena including stress sensing, stress response, and also cellular suicide (programmed cell death or **apoptosis**) as a consequence of prolonged stress and the cells' answer to it. The reason why this is so lies in the common evolutionary origin of all of the stress phenomena that exist, from the most primitive forms of life to the most highly developed. In this chapter, we want to describe stress in a general way, and we want to do so for two reasons. First, we know from brain physiology that psychological stress manifests itself as oxidative stress in the brain (Massaad et al. 2009; see also Chap. 4), and most prominently and importantly for the present topic, in neuronal cells of the **hippocampus**. Second, oxidative stress is one of the most primordial forms of stress (for evolutionary reasons to be explained below) and the cells' response to oxidative stress is highly conserved from single-celled organisms to humans. "Conserved" is a very common general term in biological literature. It is simply shorthand for exhibiting a similar sequence on both the DNA and protein levels in different presently living species. The sequences were created during evolution by the Darwinian process of mutation and selection but remained similar because the function of the protein in question is conserved and can in this case only be achieved by sequence conservation. There exist very highly conserved as well as highly variable protein sequences, depending on the details of the structure and function of the protein. The sequence determines the structure, and structure determines function in a nearly one-to-one relationship.

For several reasons we are concentrating here on the genetic systems of (oxidative) stress response in single-celled eukaryotic organisms (primarily yeast) and in cultured human cells:

- The very highly developed genetic systems of yeast and of other simple model systems allow us to gain knowledge in this field more easily than in mammalian systems (primarily the mouse); this is true for transgenic mouse research as well as for organotypic cell culture research. What the yeast and other eukaryotic studies have revealed up to now is a surprising evolutionary conservation of the stress response systems.
- The many stress responses that we can experimentally study and observe require a set of genes and of gene regulations which to a large degree are

overlapping and highly conserved and to a smaller degree are divergent between different species.

- Oxidative stress response seems to be one of the central pathways in that it is also activated in many cases by other stresses, such as, for example, heat stress and osmotic stress.
- Most importantly for our purposes, oxidative stress and oxidative stress responses also play a major role in the neural changes observed under psychological stress (Massaad and Klann 2011), which is a major topic of this book. Such neural changes are shown by numerous investigations both in human studies and in experiments on psychological stress in rats and mice. The role of reactive oxygen species (ROS) is a dual one depending on the ROS concentrations. At low concentrations, ROS (mainly hydrogen peroxide) are second messengers acting in synaptic transmission, but at higher concentrations and when chronically produced they can interfere with synaptic transmission, and thereby with memory formation and learning (Massaad and Klann 2011).

ROS and Their Function: The Basics of Oxidative Stress Defense

Reactive oxygen species (ROS) will play a major role in the following chapters. Hence, we start with explaining the term ROS, their formation, and their functions, as described by Aung-Htut et al. (2012). ROS are metabolic products of the **single electron reduction steps of molecular oxygen** (O_2), which are shown schematically in Fig. 3.1a. Transfer of a single electron to the dioxygen molecule creates superoxide, the oxygen **anion radical**, $O_2 \cdot^-$. Transfer of another single electron leads to H_2O_2 (hydrogen peroxide). The next step of reduction leads to the hydroxyl radical, $\cdot OH$. All three molecular species occur in living cells and exert distinct biological actions. Only the addition of a fourth electron yields the final reduction product, water (H_2O). H_2O_2 is a non-radical ROS that fulfills numerous biological activities as a signaling substance in all living cells, including neurons of the brain. However, depending on concentration and cellular location, it can also lead to devastating activities in the cells; a common example would be the oxidation of SH groups on proteins. The $\cdot OH$ radical is the most reactive and therefore short-lived (with a half-life on the order of nanoseconds) biological ROS that we know. It can oxidize proteins, RNA, DNA, and lipids and is toxic and mutagenic. The $\cdot OH$ radical arises under in vivo conditions through the combined

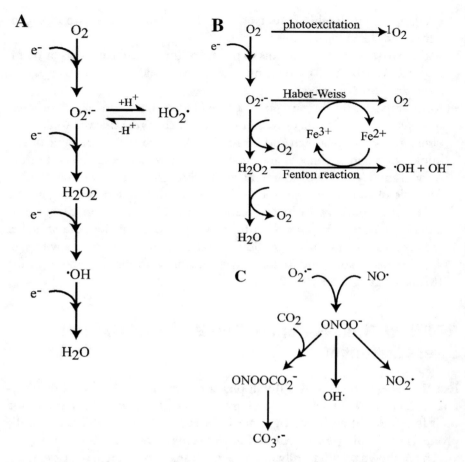

Fig. 3.1 Part a: The single-electron reduction steps of oxygen. This figure schematically summarizes the single electron reduction steps of oxygen (O_2) as they occur in living cells. The first reduction product is superoxide ($O_2\cdot^-$), an anion radical that in vivo is short-lived because of the presence of superoxide dismutase in practically all cells. Superoxide equilibrates with H^+ thereby becoming a radical that can easily pass through biological membranes. In the second reduction step, hydrogen peroxide (H_2O_2) is formed, which is not a radical but a very important reactive oxygen species (ROS). Hydrogen peroxide can be transformed into hydroxyl radicals ($\cdot OH$) (see part **b**) and only after a further reduction the final product, water, is formed. Source: Aung-Htut et al. (2012) with permission of authors and publisher. Part **b**: Scheme of the Haber–Weiss and Fenton reaction. This figure shows how the highly toxic and mutagenic hydroxyl radical is formed in living cells. This requires the cooperation of superoxide and hydrogen peroxide and free (uncomplexed) ferrous or cuprous ions. In a cyclic process depending on the valence change of the iron ions, the hydroxyl radical is formed. Source: Aung-Htut et al. (2012) with permission of authors and publisher. Part **c**: Reactions of nitric oxide. Via interaction with the biological nitrogen radical nitric oxide (NO), superoxide can also form a series of highly reactive biological radicals, among them peroxynitrite ($ONOO^-$), a major defense substance in the innate immune system. In the subsequent reaction steps shown, nitric dioxide radical and carbonate radicals are generated both of which are highly reactive and toxic. Source: Aung-Htut et al. (2012) with permission of authors and publisher

Haber–Weiss and Fenton reactions (Fig. 3.1b), which depend on free transition metal ions (e.g., Fe^{2+} or Cu^+). This is why free iron ions practically never occur in living cells but are always complexed with organic ligands. Other ROS arise by the reaction of superoxide with the biologically generated $NO^.$ nitroxyl radical (Fig. 3.1c), another biomolecule that is a radical and also fulfills essential biological actions in signal transduction. The reaction product of $NO^.$ and superoxide yields the most important and highly reactive non-radical ROS, peroxynitrite. There is now a tendency to talk about reactive oxygen and nitrogen species (**RONS**) to also cover such species as peroxynitrite. ROS fulfill important functions as signaling molecules in metabolism, yet at the same time, they are major molecules creating oxidative stress.

Oxidative stress has been a side-product of normal cellular metabolism ever since the earth's atmosphere first contained oxygen and has, since early times, been utilized for signaling and regulation in living cells. For this reason, an elaborate system of oxidative stress response and defense pathways has evolved which is necessary for survival. These pathways (Fig. 3.2) evolved for defense

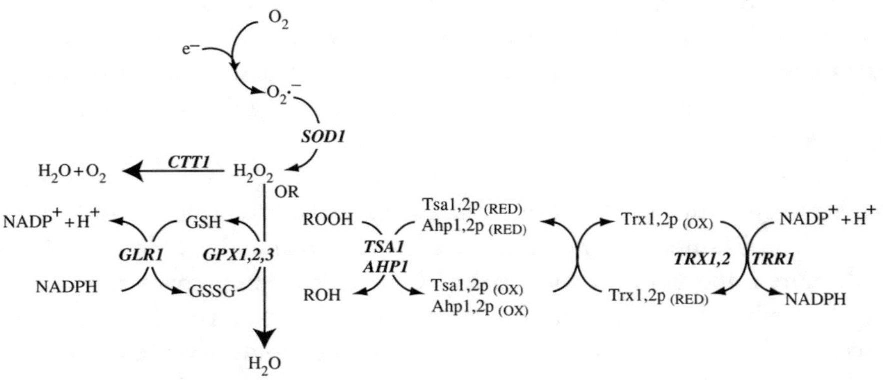

Fig. 3.2 Main pathways of detoxification of reactive oxygen species. The figure shows important examples of ROS and defense pathways in eukaryotic cells. The main reactive oxygen species are superoxide, hydrogen peroxide, and organic peroxides that are detoxified to water or to organic alcohols; detoxification occurs via the Cu, Zn-superoxide dismutase, catalase, or the two glutathione-dependent pathways, the glutathione peroxidase pathway or the thioredoxin pathway. The glutathione peroxide pathway is specific for hydrogen peroxide. The thioredoxin pathway can recognize both hydrogen peroxide and organic peroxides and reduce them to water or the corresponding alcohols, respectively. The gene symbols in this figure are the ones from yeast, the prototypic eukaryotic system: SOD1, (Cu, Zn) superoxide dismutase; CTT1, catalase T; GPX1,2,3, glutathione peroxidase isoforms 1, 2 and 3; GLR1, glutathione reductase 1; TSA1, peroxiredoxin isoform TSA1; AHP1, peroxiredoxin isoform AHP1; TRX1,2, thioredoxin isoforms 1, 2; TRR1, thioredoxin reductase 1. Source: Aung-Htut et al. (2012) with permission of authors and publisher

in all subcellular compartments of eukaryotic cells but were at the same time also used for the transduction and modification of signals in those cells (Aung-Htut et al. 2012). To illustrate this, we present here the major oxidative stress defense pathways existing in the cytosol of eukaryotic cells (Fig. 3.2).

Superoxide is in all cases the first product of oxygen one-electron reduction and the parent compound for most of the ROS to be found in the cell. It can arise either by a side reaction of the respiratory chain in **mitochondria** or by a dedicated reaction catalyzed by **NADPH oxidases**. NADPH oxidases are a group of enzymes that make superoxide "on purpose," either for specialized metabolic reactions, for defense as part of the innate immune system of higher organisms, or as a signaling substance (Breitenbach et al. 2018). In the latter case, the NADPH oxidases (for instance the human protein **Nox4**) are closely coupled in space and time with **superoxide dismutase** (**SOD**, see below), because in this way the ultimate signaling substance, which is in most cases H_2O_2, can be produced in a rapid and localized way as needed (Reddi and Culotta 2013). Superoxide is rapidly converted to H_2O_2 and oxygen by the SOD enzymes. SOD1, the copper-zinc enzyme, resides in the **cytoplasm** and in the intermembrane space of the mitochondria; SOD2, the manganese enzyme, is located in the mitochondrial matrix. A third form of SOD is an extracellular isoform of the copper-zinc SOD, and like the other two isoforms fulfills an important defense role in higher organisms (Nozik-Grayck et al. 2005).

Defense against ROS can either be achieved by direct conversion of superoxide to H_2O_2 (described above), or the excessive or mislocalized H_2O_2 can be detoxified (dismutated) by catalase, resulting in O_2 and H_2O. However, H_2O_2 and, more importantly, a wide variety of organic hydroperoxides, which are highly detrimental to cells, can be reduced by two antioxidative pathways that occur in all eukaryotic cells. These are the **glutathione peroxidase** pathway and the **thioredoxin pathway** (Fig. 3.2). Both pathways strictly depend on **NADPH** as the ultimate reducing agent. This is the reason why the **pentose phosphate pathway** (the main producer and source of NADPH) must be activated in cells under oxidative stress. Components of these two antioxidative pathways mentioned are also used to protect protein thiol groups in order to retain the enzymic activity of thiol-based enzymes (Fig. 3.3). The same components (e.g., the peroxiredoxin, Tsa1) are also in certain situations used in signal transduction pathways.

The stress response pathways that have developed over the billions of years of biological evolution are, on the one hand, diverse but interrelated and characterized by cross-reactivity. On the other hand, they contain many highly conserved signaling modules that point to a common evolutionary origin for

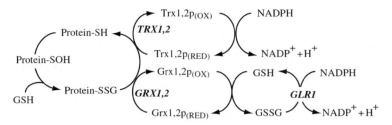

Fig. 3.3 Maintenance of protein thiol groups. This figure shows the pathways operating in the cytosol of eukaryotic cells which can maintain the reduced state of protein thiol groups. Oxidized protein thiol groups can be reduced either by the glutathione-based system consisting of glutaredoxin 1,2, glutathione reductase, glutathione, and NADPH or the thioredoxin-based system (thioredoxin 1,2, thioredoxin reductases, and NADPH). Maintaining the reduced state of protein thiol groups is essential for survival under oxidative stress. The gene symbols in this figure are the ones from yeast, the prototypic eukaryotic system: GLR1, glutathione reductase 1; TRX1,2, thioredoxin isoforms 1, 2; TRR1, thioredoxin reductase 1. Source: Aung-Htut et al. (2012) with permission of authors and publisher

these biochemical modules. We will further explain this by giving illustrative examples rather than attempting to be comprehensive.

The Integrated Stress Response and Its Relationship with Oxidative Stress

One of the most illustrative examples is the so-called integrated stress response (ISR) (Harding et al. 2003; Zhu et al. 2019; Pakos-Zebrucka et al. 2016). This pathway is highly conserved in all eukaryotic cells from yeast to humans and plays an important role in influencing synaptic transmission in the brain. A key and highly conserved central part of this pathway is the protein **eIF2alpha** (eukaryotic initiation factor 2 alpha), a regulator of protein synthesis, which can bind to the **ribosome** small subunit during the start of protein synthesis and stimulate or inhibit it, depending on its phosphorylation status on serine 51. If phosphorylated, the synthesis of the large majority of cellular proteins (those needed in growing cells) is strongly reduced at the level of **translation**. However, a small subset of proteins is now transcriptionally and translationally upregulated, which guarantees downregulation of housekeeping proteins but upregulation of proteins needed for a metabolically near-silent survival state. A key transcription factor responsible for the latter function is Atf4, which is the functional homolog of the transcription factor Gcn4 of lower (single-celled) eukaryotes. Translation of this central stress response transcription factor is regulated by eIF2alpha phosphorylation. Among the survival functions regulated by Atf4 in human cells is the synthesis of the biological antioxidant, **glutathione** (**GSH**), and the buildup of reserve nutrients that help cells to survive starvation, but also the induction of the **autophagy** pathway (Kroemer et al. 2010). Phosphorylation of eIF2alpha is performed by a multiplicity of protein kinases of the family of PKR-like ER protein kinases, among them PERK (PKR-like ER kinase), PKR (double-stranded RNA-dependent protein kinase), HRI (heme regulated eIF2alpha kinase), and GCN2 (general control non-derepressible kinase). Within the ISR system, these kinases respond to various perturbations of **homeostasis**, like oxidative stress, amino acid starvation, and in the case of PKR, virus infection. As endpoints of this stress response system, we can recognize at least three outcomes: restoration, adaptation, and resumption of growth and metabolism; cell cycle arrest and dormancy leading to survival; or the entry into the pathways of programmed cell death. Which of these endpoints comes particularly into effect depends on the starting protein kinase and the overall level and kind of stress. In the first phase of this stress response pathway, phosphorylated eIF2alpha is blocked from translation initiation of **cap**-dependent messenger RNAs (**mRNA**s), and at the same time gains the ability to recognize internal start codons independently of the 5' cap of the mRNA. The most important example is the translation of the aforementioned transcription factor Atf4. It controls genes, which are silent under normal or growth conditions but activated under stress conditions. Translation of Atf4 under unstressed conditions is blocked by the upstream short open reading frames of this mRNA; however, when eIF2alpha is phosphorylated, the genuine start codon of the Atf4 mRNA is recognized. Downstream of Atf4 in the signaling pathway, for instance, lies the activation of a transcription factor needed for apoptosis. Of note, apoptosis in multicellular organisms can have a positive survival value, as heavily damaged cells are removed and replaced by stem cell division. As already mentioned, Atf4 is the functional homolog of the Gcn4 transcription factor of lower eukaryotes.

So, eIF2alpha is a central point or hub of integration for different and overlapping kinds of stress. This central regulatory protein can obviously (as described above) exert other functions (partially still unknown in mechanistic detail) in addition to the initiation of translation on the ribosome. It can indirectly but strongly activate the transcription of a subset of genes needed under stress conditions, which are only marginally transcribed in growing cells. We can summarize that the regulation of the stress response in living cells is not linear but consists of a network structure comprising multiple stresses and using conserved cellular biochemical modules. This system is at work in all cells including, as has been pointed out previously, neurons of the brain (hippocampus) during learning and **long-term potentiation** (**LTP**), which can be accompanied by oxidative stress. The activity of the ISR in hippocampal glutamatergic neurons has been determined experimentally in the mouse. The methods employed were genetic deletion or downregulation by small inhibitory RNA constructs of ISR components in the mouse (for instance *ATF4*) and behavioral learning tests, as well as measuring synaptic potentials by microelectrodes. On the one hand, low activation of the ISR seemed to enhance long-term potentiation and learning (Pasini et al. 2015)—on the other hand, stronger or chronic activation of the ISR seemed to block synaptic plasticity and learning (Costa-Mattioli et al. 2005). This apparent contradiction is resolved when the level and chronicity of the stress are considered (discussed in Massaad and Klann 2011).

In a recent paper, it became clear that the ISR must be fine-tuned (including de-activation in the absence of stress) in healthy human cells (Zhu et al. 2019). These authors showed that inappropriate activation of the integrated stress response is responsible for the cognitive impairment in Down's syndrome.

Stress and the Origin of Life

In the next part of this chapter, we will briefly summarize the evolutionary origin of oxidative stress, thus giving a tentative explanation for the universality of the stress response mechanisms that have been found by experimental investigations.

Life originated on a completely anoxic earth. To understand what happened to early life, or the **eobionts**, as they are often called, one must consider how atmospheric oxygen originated on earth. All of the atmospheric oxygen present during evolution originated from photosynthesis (see below). Life as we know it depends on a small number of necessary metabolic prerequisites. Among the many origin-of-life theories that exist, we think that the "metabolism first" theory (Wächtershäuser 1990) is the most attractive. This theory places emphasis on the hypothesis that primitive metabolic cycles pre-dated the formation of the first cells. Energy requirements had to be met by the appropriate metabolic pathways to maintain the structure and function of living cells.

What are the most important prerequisites that must be fulfilled if cells are to live? Cells are alive only if they are enclosed by a membrane not freely permeable to ions like sodium, potassium, chloride, calcium, and many others.

The cells must have an energy source, and they must have metabolism, using an energy source to break down energy-rich molecules to produce end products that have a lower energy content than the starting substances. Metabolism thus generates a chemical difference between the inside of the cell and its outside environment, and thereby an electrochemical potential gradient across the cellular membrane which normally is about 280–310 mV with the inside of the cell being negatively charged. Any physical action (like a hole in the membrane) that removes this potential gradient kills the cell.

In addition to the energy requirement, there must be a mode to grow and to multiply—otherwise, the cells would have been eliminated by Darwinian evolution a long time ago. Furthermore, there must be a mechanism guaranteeing relative genetic stability. This means that the cells must have a genome, which is a chemical entity that remains nearly constant and changes very little over many cell generations. DNA is such a molecule. On the other hand, very small changes in DNA are necessary to cause molecular evolution by mutation, thereby providing the basis for variation and selection as first described by Charles Darwin (1859). The genome must be chemically much more stable than the rest of the biomolecules (this idea was already developed in Schrödinger's famous little book, *What Is Life?*, first published in 1944). Today this idea has to be modified: DNA is a chemically stable molecule, but what is more important for the genetic stability of the dividing and growing cells is the incredibly elaborate system of **DNA repair** which was completely unknown at the time of Schrödinger's book. The discovery of DNA repair (photo repair) dates to 1949 (Kelner 1949), but many of the most important pathways of DNA repair known today were only discovered decades later (Friedberg et al. 2006). Whether the first genomes were RNA, for which in the modern world practically no repair systems exist, or DNA, we do not know. However, we emphasize that the arguments made here about essential features of living cells remain strong irrespective of whether the popular hypothesis of an "RNA world" (Gilbert 1986) is correct.

Stress and Molecular Evolution

Genetic stability is necessary for survival, because only genetic stability can lead to the continued production of the same biomolecules, including most importantly metabolic enzymes, which can guarantee the stability of metabolism. However, in reality, the genomic sequences of living cells spontaneously change by mutation (Muller 1927; De Vries 1901). The reasons for (or biochemical mechanisms behind) spontaneous mutations were at first unknown and debated intensively and were only discovered many decades later (for review: Griffith et al. 1993). The actually observed spontaneous mutation frequency in all biological systems, eukaryotic and prokaryotic, has been found to be between one in a billion and one in ten billion, meaning one fixed mutation in the course of 10^9 or 10^{10} nucleotides replicated (Drake 1991; Luria and Delbrück 1943; Kondrashov and Kondrashov 2010). To put this in context, we must consider the number of base pairs (bp) in the (haploid) human genome: about 3.5×10^9. This frequency seems to be an optimum: it is sufficiently small to allow genetic stability, and it is sufficiently large to allow biological evolution, which is based on sequence changes in DNA and usually takes a large number of generations.

Of course, this is true only in a world where the genome both carries and store information. It is often called a blueprint for the construction of gene products, and encoded in the genome are the workhorses of living cells, meaning the metabolic enzymes and other biomacromolecules like ribonucleic acids (RNA) in the form of mRNA, structural RNA, ribozymes, and numerous other regulatory RNAs. All those gene products are encoded by a relatively small part of the genome, and in addition, in the larger "non-coding" part of the genome, a large and growing number of regulatory sequences are found, which do not produce a gene product but are also part of the genome.

In the first, anaerobic, phase of life on this earth, the energy source for life was chemical energy. Energy-rich compounds were originally formed abiotically in the oceans of the primordial earth but were eventually used up. One successful additional source of biological energy resulted from oxygenic photosynthesis, utilizing the energy of sunlight. Oxygenic photosynthesis was invented by eobionts, which were similar to the contemporary blue-green bacteria (cyanobacteria). The term "invented" here needs an explanation for the non-specialist. What we mean by "invented" is that the process in question came into being by mutation and selection and for that reason served a purpose and could survive in evolution.

Oxygenic photosynthesis was invented only once, and the origin of chloroplasts in eukaryotes is monophyletic—all living cells capable of oxygenic photosynthesis are now believed with strong arguments to be derived from these original photosynthetic prokaryotes (Rodríguez-Ezpeleta et al. 2005). In fact, all the oxygen that now exists in the atmosphere results from oxygenic photosynthesis. However, oxygen was stressful and toxic to the early living cells, (and still is toxic to living cells, depending on its concentration) (Clark et al. 2006). With the advent of oxygen, living cells first evolved enzymatic mechanisms for detoxifying oxygen. These early detoxifying enzymes were capable of using oxygen for metabolism and are presumably the precursors of the

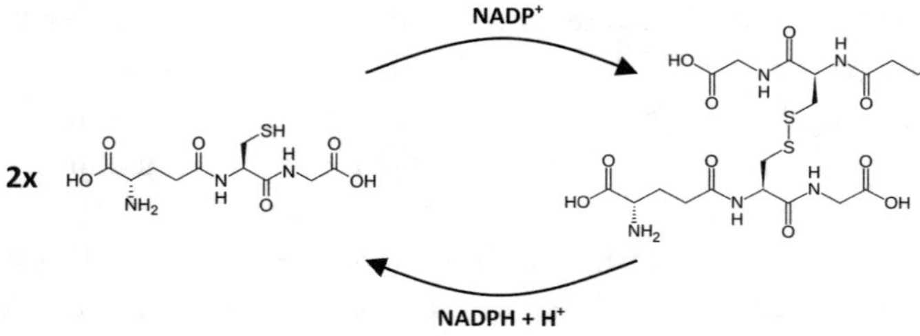

Fig. 3.4 Structure and redox reactions of the glutathione system. In the left part of the figure, the structure of the iso-tripeptide, glutathione, is shown. Two molecules of glutathione (GSH) are oxidized by NADP+ to one molecule of the disulfide form, GSSG. This reaction is reversible. Catalyzed by glutathione reductase and NADPH + H+, GSSG is reduced back to GSH. This redox reaction occurs in all living cells and is essential for survival

present-day enzymes of oxidative stress defense. Many of them remained functionally unchanged through millions of years of evolution, as evidenced by their occurrence in widely diverged modern organisms like yeast and humans and their sequence similarity (compare the examples given in Figs. 3.2, 3.3, and 3.4). However, to a lesser degree, oxidative stress defense mechanisms, although present in all aerobic cells, have also diversified through evolutionary history, in response to environmental conditions (Siauciunaite et al. 2019). Arguably, the detoxification mechanisms that evolved were also the origin of the eukaryotic pathway of respiration, which originated mono-phyletically and is located in mitochondria of nearly all eukaryotic cells, enabling those cells to create chemical energy in the form of adenosine tri-phosphate (**ATP**) in a much more efficient way than by fermentative metabo-lism (where metabolism creates ATP and uses chemical energy without the involvement of oxygen) alone. It has been speculated that the creation of multicellular organisms was only possible after this more efficient way of energy metabolism arose (Lane 2005). But the toxicity of oxygen remains even in the modern world and with it the necessity to defend against oxida-tive stress.

What Is Oxidative Stress in Cellular or Biochemical Terms?

Because this is such a central term in this book, we want to spend some time clarifying this question and will compare three different ways of defining and understanding oxidative stress. The first definition is, in very simple terms, the preponderance of the oxidized forms of biomolecules over their corresponding reduced forms. To explain this, we will first deal with the **redox reactions** that take place in living cells. Redox reactions are all those chemical reactions in which one of the two reaction partners is oxidized and the other one is accordingly reduced. Redox reactions play central roles in cellular metabolism. Every pair of oxidized and reduced biomolecules forms a **redox couple**. A common example is the redox couple consisting of oxidized and reduced glutathione (Fig. 3.4 and see below). Metabolic reactions can dynamically interchange the two forms of a redox couple, thereby using up and also transporting the most typical redox equivalents of the cell. Within the limits given by the electrochemical potential of the cell, which is reducing inside of the cell, many biological redox couples are based on sulfhydryl (SH) groups which represent the reduced forms of these redox couples, while the oxidized forms are molecules containing the corresponding disulfide bridges. The most prevalent "redox buffer" in all cells is the glutathione (GSH)—oxidized glutathione (GSSG) couple, with the equilibrium lying very far on the side of the reduced form under normal physiological conditions. This includes neurons and **astrocytes** in the brain (McBean 2017; Aoyama and Nakaki 2015). The primary source of oxidation equivalents involved in the interconnected redox equilibria is O_2 (molecular oxygen), while in most cases the reduction equivalents are **NADH** and NADPH. NADPH is vitally necessary for oxidative stress defense and is primarily created through the pentose phosphate pathway (Stincone et al. 2015). NADPH is also needed for many anabolic reactions, meaning reactions that build up cellular material and thereby increase biomass for newly formed cells (Wamelink et al. 2008). How oxygen, on the one hand, and NADPH on the other, are integrated into the redox metabolism of the cell is discussed in more detail below. If, as mentioned above, the ratio of reduced versus oxidized forms of the redox couples in the cell are shifted toward the oxidized form (GSSG in the case of the glutathione redox couple), we speak of oxidative stress.

The second definition of oxidative stress takes into account the concentrations of all of the redox-active substances in their oxidized and reduced forms. This would theoretically allow us to calculate in quantitative physicochemical terms the amount of the oxidative stress given as a deviation from the physiological negative potential gradient across the plasma membrane of the cell. However, this definition is not useful in practice, because the necessary values for the redox concentrations are difficult to obtain, especially if one considers that different subcellular organelles display different values of the redox potential. To give one example, the redox potential of the cytoplasm is around −310 mV, but the corresponding value of the lumen of the **endoplasmic reticulum**

(**ER**) is considerably more positive, enabling the formation of disulfide bridges in proteins to be secreted from the cell, which takes place in the ER. Moreover, the overall values obtained for the redox metabolites of a cell by methods of metabolomics (nowadays mostly highly developed mass spectrometric analysis) do not reflect what is relevant locally in the cell.

Therefore, a third definition of oxidative stress is preferred because of its practical usefulness: "The situation when due to some reasons the steady state ROS concentration is acutely or chronically increased leading to oxidative modification of cellular constituents resulting in disturbance of cellular metabolism and regulatory pathways, particularly ROS-based, has been called oxidative stress" (Lushchak 2011; see also Lushchak 2014). This definition is practically useful, because methods have been developed for quantitatively analyzing ROS and their follow-up products in stressed cells, enabling the estimation of the amount of oxidative stress.

Next, we will describe how living cells handle oxidative stress by adjusting their metabolism and restoring the healthy balance of oxidizing and reducing molecules.

ROS Metabolism and Antioxidative Stress-Response Pathways in Cells

Most ROS (reactive oxygen species) and **RNS** (**reactive nitrogen species**), collectively called RONS, arise ultimately from the one-electron reduction of O_2, leading to the formation of the superoxide radical anion (Fig. 3.1a). We will now discuss a few selected and pathophysiologically most important molecules derived from superoxide and the products of ROS generated in their reaction with DNA, lipids, and proteins. All living cells contain one or more superoxide dismutases (Sod), which convert superoxide in a **disproportionation reaction** to hydrogen peroxide and oxygen. Hydrogen peroxide (H_2O_2) is not a radical but by far the most important oxidizing species in living cells and will be treated in more detail below. The already mentioned Fenton and Haber–Weiss reactions (Fig. 3.1b) in the cell lead to a cyclic process converting H_2O_2 to the ·OH Radical, which is so highly reactive that it will react in a diffusion-controlled way with practically all biomolecules encountered—proteins, nucleic acids, lipids, and many low-molecular-weight metabolites (Halliwell and Gutteridge 2007). The reaction requires catalytic-free (uncomplexed) ferrous or cuprous ions. As both transition metals are trace elements absolutely necessary for life, Nature has developed an elaborate protection system to take up and metabolize these metal ions without taking the risk of inducing the Fenton reaction. Some of the reaction products formed in cells containing ·OH radicals are highly toxic and/or mutagenic. One prominent example is **HNE (4-hydroxynonenal)**, a powerful biological mutagen (Schaur et al. 2015). Other prominent products of cellular metabolism under conditions of oxidative stress induce defense reactions that are useful in providing oxidative stress markers because of their relative chemical stability, which is in contrast to the short-lived ROS molecules. Among them are protein carbonyls, lipid hydroperoxides, and low-molecular-weight substances like the so-called TBARS (thiobarbituric acid reactive substances) that result from the reaction of ROS. The prototypic example is malondialdehyde and related aldehydic substances. These markers of oxidative stress are less extremely toxic but more stable than HNE and are very useful for estimating the degree of oxidative stress prevalent in the cells, thereby integrating the amount of oxidative stress suffered over time. An important example is the accumulation of such reaction products under oxidative stress in the serum of patients suffering from neurodegenerative or psychiatric diseases and in mouse models of psychological stress to be described later (Tsai and Huang 2015).

The oxidatively derivatized proteins have a strong tendency to lose their native structure, to become enzymatically inactive, and to aggregate. Such aggregations are a feature of living cells even in the absence of oxidative stress, although the spontaneous tendency to protein aggregation is significantly increased in stressed cells. Therefore, Nature has developed an elaborate system to counteract this tendency, consisting of **chaperonins** and **chaperones**, which can assist the refolding of denatured proteins (Yang et al. 2019); degradative systems like the proteasome (Prakash and Matouschek 2004) and autophagy (Ohsumi 2014; Meng et al. 2019; Sánchez-Martín and Komatsu 2019), which can recognize denatured proteins; and ultimately also the asymmetric distribution of misfolded proteins between rejuvenated and old cells during somatic cell cycles (Klinger et al. 2010). Autophagy, proteasomal degradation, and apoptosis play major roles in oxidative stress in the brain, and they interact with each other, for instance in the form of chaperone-mediated autophagy (Alfaro et al. 2019; Kroemer et al. 2010).

The reaction of superoxide with NO· (nitric oxide, a radical and important second messenger in the cell) produces peroxynitrite, a non-radical, highly aggressive molecule likewise reactive with proteins, lipids, and nucleic acids. Peroxynitrite can disintegrate to yield ·OH radicals, and also nitroxyl radicals, which may lead to nitro compounds (e.g., nitrotyrosine) that are stable, strong indicators of oxidative stress (Verrastro et al. 2015). Peroxynitrite is also part of the inborn non-adaptive immune defense, the first line of defense against invading pathogens. The relevant chemistry is induced in macrophages and monocytes (Breitenbach et al. 2018; Thomas 2017) and will not be described in detail in the context of the present book.

The biochemical examples given above show that oxidative stress and the defense against it lead to a variety of changes, in particular in neurons of the brain, which have a pronounced effect on brain functions like LTP and learning and are part of psychological stress, which we deal with in Chap. 4.

The Role of Hydrogen Peroxide

Hydrogen peroxide is one of the most remarkable and biologically important oxidants, and its known functions merit description. To start with, it is the only oxidant for which a signaling function has been analyzed and published extensively. H_2O_2 is produced by normal cellular metabolism. Examples are fatty acid desaturases of peroxisomes; the *ERO1–PDI1* protein disulfide oxidation system of the endoplasmic reticulum, which uses O_2 as terminal acceptor and produces H_2O_2 (Zito 2015); and, in eukaryotes, the family of NADPH oxidases in conjunction with superoxide dismutases that produce H_2O_2 not only for defense but also as an important cellular signaling substance (Breitenbach et al. 2018), as already mentioned. H_2O_2 is also generated by accident, originating from a "leaky" mitochondrial respiratory chain via superoxide and superoxide dismutase (Chance et al. 1979). It is a neutral, non-radical, reasonably stable molecule that can enter cells easily, presumably via aquaporins (Bienert and Chaumont 2014). It is therefore also important for understanding ecological interactions between different species of free-living microorganisms. Because H_2O_2 in the cell is highly toxic, a number of detoxifying pathways have developed. The catalase system is specific for H_2O_2, converting it directly to O_2 and H_2O. This is a simple disproportionation reaction that is not based on reduction equivalents like NADPH. The second enzymatic system dealing with H_2O_2 is based on glutathione peroxidase, linked to glutathione, glutathione reductase, and NADPH, and specific for H_2O_2. The third and most important cellular antioxidative system is based on the two-cysteine-peroxiredoxins, which in many cells are the most abundant proteins, reaching concentrations of up to 1% of the soluble protein mass of the cell. Peroxiredoxins are highly conserved in sequence and function from bacterial to human cells, are often encoded by several different genes, and can reduce hydroperoxides like H_2O_2 and the extremely important and toxic alkyl- and fatty acid-derived lipid hydroperoxides formed under oxidative stress. The system ultimately depends on NADPH as a reductant and is involved not only in detoxification but also in signal transduction. The reaction cycle of peroxiredoxins has been elucidated in great detail, involving the intermediary formation of sulfenic acid from active site cysteine SH groups of the redox enzymes, the formation of the corresponding alkyl alcohol which is usually non-toxic, and water as the ultimate reaction products (Fig. 3.2). The oxidation equivalents are then further metabolized via thioredoxin and ultimately NADPH. Under very strong oxidative stress, which is often accompanied by the unfolding of proteins, eukaryotic (but not prokaryotic) peroxiredoxins are further oxidized to the sulfinic acid form, then become inactive as redox enzymes, but powerful as molecular chaperones that are needed in such a situation to avoid excessive aggregation of the denatured proteins. If the amount of protein denaturation and clustering is no longer large, the sulfinic acid form of peroxiredoxin is reduced back by sulfiredoxin and can again take part in the redox cycle.

Stress and the Subcellular Compartmentation of Eukaryotic Cells

Eukaryotic microbial systems are confronted with the problem of dealing with H_2O_2 in their environment. Both H_2O_2 and organic peroxides are produced by regular aerobic metabolism and are used as a means to communicate with other microorganisms (Marinho et al. 2014). However, two new features of eukaryotic life come into play in comparison with prokaryotes. The presence of a rich subcellular compartmentation creating different isolated subcellular spaces, in particular mitochondria, the ER, the nucleus, among others, enables regulation of **gene expression** by shuttling regulators (for instance, transcription factors) between nucleus and cytoplasm and also makes it possible for high local concentrations needed for signaling in some cases to be maintained without compromising the survival of the cell. Furthermore, NADPH oxidases (so-called Nox enzymes) were developed in eukaryotes by evolution to serve the sole purpose of the production of superoxide radical anions and are often coupled spatially with superoxide dismutases (Sod enzymes), which convert superoxide to H_2O_2 (Rinnerthaler et al. 2012; Leadsham et al. 2013; Breitenbach et al. 2018). Recently, a cyanobacterial Nox enzyme was discovered (Magnani et al. 2017). At present, it is unclear if this prokaryotic enzyme was imported from eukaryotic into cyanobacterial cells during evolution. Of course, the second purpose of the oxidant production by Nox enzymes is defense, and in some special cases the performance of specialized metabolic reactions (Breitenbach et al. 2018). In the majority of cases, H_2O_2 is the signaling molecule, while only in exceptional cases is the signal transmitted directly by the superoxide radical anion (Yang and Hekimi 2010). Indirect genetic evidence using the eukaryotic model organism, *S. cerevisiae*, clearly shows that oxidative stress in various forms, if applied in the medium, can activate a transcriptional response of the target genes mediated by a number of stress response transcription factors. In the yeast system, a detailed molecular mechanism of gene expression during the stress response has been elucidated involving translational regulation. This is the integrative stress response described above, which occurs universally in eukaryotic cells.

Oxidative Stress Sensors of Eukaryotic Cells and Transmission of the Oxidative Stress Signal

Many mechanistic details of oxidative stress signaling are still unknown, including identification of the precise cysteine SH groups that need to be oxidized for signal transmission. This would require experiments with purified components and using an in vitro system to measure gene expression at the mRNA and protein level. Such a detailed understanding is presently only available in exceptional cases, which include the yeast oxidative stress transcription factor, Yap1 (yeast activating protein 1, Marinho et al. 2014), as well as in the mammalian oxidative stress signal transmission described below (Chen et al. 2008, 2009). However, what the yeast genetic experiments do show rather clearly is that there is cross-activation of other stress systems, like the heat shock response, when oxidative stress is applied. A primary example is the induction of the gene expression signature of oxidative stress mediated by the heat shock transcription factor, Hsf1 (Yamamoto et al. 2007). Marinho et al. (2014) give an overview of which yeast transcriptional control systems are involved and strongly responsive to external oxidative stress. They are: Yap1, Maf1, Hsf1, Msn2, and Msn4. Importantly, the yeast essential gene, *HSF1*, is conserved structurally and functionally in mammalian cells indicating the universality of the oxidative and heat stress response. The human cDNA can complement the yeast null mutant. Transcription of target genes of Hsf1 under oxidative stress is mediated by the oxidants in an unknown way, resulting in Hsf1 phosphorylation and nuclear transfer (Yamamoto et al. 2007; Taymaz-Nikerel et al. 2016).

The second highly conserved eukaryotic transcription factor family involved in stress response is AP-1 (activating protein 1), in yeast oxidative stress research mainly represented by the nonessential yeast gene, *YAP1*. The yeast members of this transcription factor family were discovered by genome sequencing and by the fact that the yeast Yap1 protein binds to a promoter element of the human virus SV40 (Toone and Jones 1999). Upstream binding elements regulating the expression of the target genes of AP-1-like factors are conserved between yeast and human genes. Yap1 is a sensor of oxidative stress which is regulated by the oxidation of two cysteine SH groups (C303, C598). The disulfide form of this transcription factor constitutes the inactivation of a nuclear export signal, thus preventing nuclear export and increasing transcription of the stress-induced target genes. Regulation through shuttling between cytoplasm and the nucleus is a common feature of eukaryotic transcriptional regulation. Oxidation of Yap1 is accomplished by the glutathione peroxidase and peroxiredoxin systems (Fig. 3.2). Downregulation of the signal created by Yap1 is mediated by calcineurin and involves dephosphorylation and proteasomal degradation (Yokoyama et al. 2006).

The target genes of the transcription factor Yap1 include many of the most important proteins of the antioxidative response, such as catalase T, glutathione peroxidase, thioredoxin reductase, and glutathione biosynthesis enzymes, but also genes involved in resistance to heavy metals, such as cadmium. In the case of oxidative stress regulation by yeast Yap1, we find a beautifully consistent picture of gene regulation: overexpression of Yap1 leads to an increase, and deletion of this nonessential transcription factor leads to a severe decrease in expression of the oxidative stress defense proteins, including many of the redox system components shown in Fig. 3.2. For this reason, we have discussed Yap1 in depth here.

However, the study of the cellular stress response pathways has not always produced results that are easy to understand. The results of "**omics**" (transcriptomics and in the last few years increasingly proteomics and metabolomics) are not always consistent compared with genome-wide phenotypic analyses of deletion mutants. The functions defined by the two methods are apparently slightly different. Gene expression studies point to genes that need to be induced for survival, whereas the deletion mutant data is more likely to represent constitutive functions needed to provide resistance. The reason for this discrepancy is to be found in the complexity of the genetic networks of the cells, even of the "simple" yeast cells; in differences in the genetic background of the often genetically different laboratory strains of yeast; and in the large frequency of unwanted and unknown suppressor mutations in the deletion strains contained in the commercially available deletion mutant collections (Winzeler et al. 1999; Thorpe et al. 2004; Teng et al. 2013). A similar problem appears in cells of different individual humans because of their enormous variety in gene expression and gene interaction despite the principal genetic unity of the global human population. To illustrate this, consider individual differences in the response to the efficacy of many pharmaceutical molecules.

Nevertheless, we wish to close this chapter with a general consideration of available methods that in combination can bring us closer to understanding oxidative stress responses and the general stress response pathways of cells.

Universality of the Stress Phenomenon: Outlook on Methods and Results

There are two major and several minor genome-wide methods for the exploration of the genetic structure of the (oxidative) stress response. All of them are used in the elucidation of stress response systems in model organisms, in transgenic mice as well as in organotypic cell culture. The list of methods below shows that stress research is part of a global attempt to understand the genome by functionally annotating every gene and every gene-regulatory sequence in the genome.

(i) The first method is measuring genome-wide transcriptional activity, which is triggered by oxidative stress, normoxia, and hypoxia. From the differences between these oxygen-dependent transcriptomes, we can draw conclusions about the response of the cells to oxygen or the lack of oxygen. The method was pioneered by DeRisi et al. (1997).

(ii) Another method is to monitor the oxidative stress resistance or sensitivity of every possible deletion mutation in the genome (Thorpe et al. 2004; Winzeler et al. 1999) and every possible overexpression construct corresponding to all genes in the genome (Sopko et al. 2006). Two specialized mutant strain collections of S. cerevisiae, which make such monitoring possible, are commercially available.

(iii) Protein-protein interactions have been systematically explored by robotic methods (Salwinski et al. 2004). The rationale is to assign function to a protein and gene by showing its interaction partners.

(iv) Likewise, synthetic lethality can help to explore the stress-related function of a gene (Tong et al. 2004).

(v) If the function in stress response of a transcription factor is known (a prominent example is Yap1), then identification of promoter binding sites by computational methods and by experimental methods can also help to identify a stress-related function of a gene (Ren et al. 2000).

A few remarks are necessary to explain how these methods can work together (Breitenbach et al. 2007):

Although these methods are most highly developed in yeast, there are a growing number of model organisms for which the same methods have been developed, namely strain collections and mutant collections corresponding to all known genes; genome sequencing and identification of all functional genes; assignment of functions to all genes (annotation); and creation of microarrays for robotic and automatized measurement of the transcription level for all genes (the transcriptome), under as many circumstances as we wish, primarily after application of (oxidative) stress. It has been necessary to develop new bioinformatics methods to be able to analyze the immense amount of primary data obtained by these methods.

Only an intelligent combination of these methods can begin to answer the relevant questions pertaining to the biology of oxidative stress responses. It became clear rather quickly that transcriptional expression of a gene and its importance for the process involved do very often go hand in hand, but not in 100% of the cases. The reasons for these problems are many; however, the main reasons are functional redundancy of the genes contained in the genome, and pleiotropic gene action, meaning a network structure in which the gene in question can be involved in several branches of the network, sometimes not obviously functionally related.

In this chapter, we have repeatedly distinguished between mild and severe oxidative stress conditions and emphasized the differences in outcome between these superficially similar situations. Both conditions lead to a change in the homeostatic dynamic state of the cell, often called "homeodynamics."

"Mild stress" leads to a new but viable homeodynamic state, while "extreme stress" can either kill the cell directly or create such a strong disturbance in cellular metabolism that the cell enters a pathway of programmed cell death. When we look at the biological oxidant H_2O_2, for example, which we already have discussed above, we see that H_2O_2 can be used to create oxidative stress in the cell, which can lead to distinct outcomes.

Up to a certain concentration (the exact amount depends on the cell type and metabolic situation), the cell reaches a viable new homeodynamic state (Klinger et al. 2010), which can, however, confer defects in certain physiological specialized reactions (e.g., cell differentiation in the form of sporulation). Above a certain limit or a certain amount of oxidative stress, the cell can enter a state of transient quiescence or it can activate one of the programs of programmed cell death (apoptosis, ferroptosis, programmed necrosis, or others), and the cell dies.

Mild stress, as we have seen, carries many similarities to "**eustress**," a term coined by Selye (Szabo et al. 2012). It activates genetic stress response programs, which among other things can increase resistance to stress. This not only applies to the same stress that was involved in the first place, but also to other stresses, and for a prolonged time (Temple et al. 2005). The biochemical basis for resistance can in this case lie in the presence of an increased amount of cell antioxidants (for instance, glutathione (Fig. 3.4), a tripeptide that is the most important "currency" of antioxidation in cells). According to our working hypothesis (see Chap. 4), this kind of stress resistance is at least in part the biochemical basis of physiological and also psychological **resilience**. In cell biology, this well-known effect is called "**hormesis**" (Rattan 2008).

On the other hand, destructive stress leading to programmed (or direct) cell death is similar to Selye's "**distress**." In medical terms, distress causes a whole number of diseases, among them the typical psychosomatic diseases, such as many forms of heart disease, where lifestyle factors and factors of psychological stress interact (see Fig. 4.1 in Chap. 4), and the broad syndrome of **major depression**, to be discussed later. In many cases, distress also works via apoptotic killing of cells, in the hippocampus of the central nervous system, for instance. It will be our aim in the next chapter to show evidence and to propose a hypothesis for how psychological stress via biochemical stress in neurons can cause a number of disease states, including depression and cognitive decline in humans and experimental animals (Massaad and Klann 2011).

References

Alfaro IE, Albornoz A, Molina A, Moreno J, Cordero K, Criollo A, Budini M (2019) Chaperone mediated autophagy in the crosstalk of neurodegenerative diseases and metabolic disorders. Front Endocrinol 9:778. https://doi.org/10.3389/fendo.2018.00778

Aoyama K, Nakaki T (2015) Glutathione in cellular redox homeostasis: association with the excitatory amino acid carrier 1 (EAAC1). Molecules 20:8742–8758. https://doi.org/10.3390/molecules20058742

Aung-Htut MT, Ayer A, Breitenbach M, Dawes IW (2012) Oxidative stresses and ageing. In: Breitenbach M et al (eds) Aging research in yeast, Subcellular biochemistry, vol 57. Springer, Dordrecht, pp 13–54. https://doi.org/10.1007/978-94-007-2561-4_2

Bienert GP, Chaumont F (2014) Aquaporin-facilitated transmembrane diffusion of hydrogen peroxide. Biochim Biophys Acta 1840:1596–1604. https://doi. org/10.1016/j.bbagen.2013.09.017

Breitenbach M, Dickinson JR, Laun P (2007) Smart genetic screens. Methods Microbiol 36:331–367. https://doi.org/10.1016/S0580-9517(06)36015-1

Breitenbach M, Rinnerthaler M, Weber M, Breitenbach-Koller H, Karl T, Cullen P, Basu S, Haskova D, Hasek J (2018) The defense and signaling role of NADPH oxidases in eukaryotic cells. Wiener Medizinische Wochenschrift 168:286–299. https://doi.org/10.1007/s10354-018-0640-4

Chance B, Sies H, Boveris A (1979) Hydroperoxide metabolism in mammalian organs. Physiol Rev 59(3):527–605. https://doi.org/10.1152/physrev.1979.59.3.527

Chen K, Kirber MT, Xiao H, Yang Y, Keaney JF Jr (2008) Regulation of ROS signal transduction by NADPH oxidase 4 localization. J Cell Biol 181(7):1129–1139. https://doi.org/10.1083/jcb.200709049

Chen K, Craige SE, Keaney JF Jr (2009) Downstream targets and intracellular compartmentalization in Nox signaling. Antioxid Redox Signal 11(10):2467–2480. https://doi.org/10.1089/ars.2009.2594

Clark JM, Lambertsen CJ, Gelfand R, Troxel AB (2006) Optimization of oxygen tolerance extension in rats by intermittent exposure. J Appl Physiol 100(3):869–879. https://doi.org/10.1152/japplphysiol.00047.2005

Costa-Mattioli M, Gobert D, Harding H, Herdy B, Azzi M, Bruno M, Bidinosti M, Mamou CB, Marcinkiewicz E, Yoshida M, Imataka H, Cuello AC, Seidah N, Sossin W, Lacaille J-C, Ron D, Nader K, Sonenberg N (2005) Translational control of hippocampal synaptic plasticity and memory by the eIF2α kinase GCN2. Nature 436:1166–1170. https://doi.org/10.1038/nature03897

Darwin C (1859) On the origin of species by means of natural selection. J. Murray, London

DeRisi JL, Iyer VR, Brown PO (1997) Exploring the metabolic and genetic control of gene expression on a genomic scale. Science 278:680–686. https://doi.org/10.1126/science.278.5338.680

De Vries H (1901) Die Mutationstheorie: Versuche und Beobachtungen über die Entstehung von Arten im Pflanzenreich. Veit, Leipzig. https://doi.org/10.5962/bhl.title.11336

Drake JW (1991) A constant rate of spontaneous mutation in DNA-based microbes. Proc Natl Acad Sci USA 88:7160–7164. https://doi.org/10.1073/pnas.88.16.7160

Friedberg E et al (2006) DNA repair and mutagenesis. ASM, Washington, DC

Gilbert W (1986) The RNA world. Nature 319:618. https://doi.org/10.1038/319618a0

Griffith A et al (1993) An introduction to genetic analysis. Freeman, New York

Halliwell B, Gutteridge JMC (2007) Free radicals in biology and medicine, 4th edn. Oxford University Press, Oxford

Harding HP, Zhang Y, Zeng H, Novoa I, Lu PD, Calfon M, Sadri N, Yun C, Popko B, Paules R, Stojdl DF, Bell JC, Hettmann T, Leiden JM, Ron D (2003) An integrated stress response regulates amino acid metabolism and resistance to oxidative stress. Mol Cell 11:619–633

Kelner A (1949) Effect of visible light on the recovery of Streptomyces griseus conidia from ultra-violet irradiation injury. Proc Natl Acad Sci USA 35(2):73–79. https://doi.org/10.1073/pnas.35.2.73

Klinger H, Rinnerthaler M, Lam YT, Laun P, Heeren G, Klocker A, Simon-Nobbe B, Dickinson JR, Dawes IW, Breitenbach M (2010) Quantitation of (a)symmetric inheritance of functional and of oxidatively damaged mitochondrial aconitase in the cell division of old yeast mother cells. Exp Gerontol 45:533–542. https://doi.org/10.1016/j.exger.2010.03.016

Kondrashov FA, Kondrashov AS (2010) Measurement of spontaneous rates of mutations in the recent past and the near future. Philos Trans R Soc B 365:1196–1176. https://doi.org/10.1098/rstb.2009.0286

Kroemer G, Mariño G, Levine B (2010) Autophagy and the integrated stress response. Mol Cell 40(2):280–293. https://doi.org/10.1016/j.molcel.2010.09.023

Lane N (2005) Power, sex, suicide: mitochondria and the meaning of life. Oxford University Press, Oxford

Leadsham JE, Sanders G, Giannaki S, Bastow EL, Hutton R, Naeimi WR, Breitenbach M, Gourlay CW (2013) Loss of cytochrome c oxidase promotes RAS-dependent ROS production from the ER resident NADPH oxidase, Yno1p, in yeast. Cell Metab 18:279–286. https://doi.org/10.1016/j.cmet.2013.07.005

Luria SE, Delbrück M (1943) Mutations of bacteria from virus sensitivity to virus resistance. Genetics 28:491–511

Lushchak VI (2011) Adaptive response to oxidative stress: bacteria, fungi, plants and animals. Comp Biochem Physiol Part C 153:175–190. https://doi.org/10.1016/j.cbpc.2010.10.004

Lushchak VI (2014) Free radicals, reactive oxygen species, oxidative stress and its classification. Chem Biol Interact 224:164–175. https://doi.org/10.1016/j.cbi.2014.10.016

Magnani F, Nenci S, Fananas EM, Ceccon M, Romero E, Fraaije MW, Mattevi A (2017) Crystal structures and atomic model of NADPH oxidase. PNAS 114(26):6764–6769. https://doi.org/10.1073/pnas.1702293114

Marinho HS, Real C, Cyrne L, Soares H, Antunes F (2014) Hydrogen peroxide sensing, signaling and regulation of transcription factors. Redox Biol 2:535–562. https://doi.org/10.1016/j.redox.2014.02.006

Massaad CA, Klann E (2011) Reactive oxygen species in the regulation of synaptic plasticity and memory. Antioxid Redox Signal 14(10):2013–2054. https://doi.org/10.1089/ars.2010.3208

Massaad CA, Washington TM, Pautler RG, Klann E (2009) Overexpression of SOD-2 reduces hippocampal superoxide and prevents memory deficits in a mouse model of Alzheimer's disease. PNAS 106(32):13576–13581. https://doi.org/10.1073/pnas.0902714106

McBean GJ (2017) Cysteine, glutathione, and thiol redox balance in astrocytes. Antioxidants 6:62. https://doi.org/10.3390/antiox6030062

Meng T, Lin S, Zhuang H, Huang H, He Z, Hu Y, Gong Q, Feng D (2019) Recent progress in the role of autophagy in neurological diseases. Cell Stress 3(5):141–161. https://doi.org/10.15698/cst2019.05.186

Muller HJ (1927) Artificial transmutation of the gene. Science 66:84–87. https://doi.org/10.1126/science.66.1699.84

Nozik-Grayck E, Suliman HB, Piantadosi CA (2005) Extracellular superoxide dismutase. Int J Biochem Cell Biol 37:2466–2471

Ohsumi Y (2014) Historical landmarks of autophagy research. Cell Res 24:9–23. https://doi.org/10.1038/cr.2013.169

Pakos-Zebrucka K, Korygal I, Mnich K, Ljujic M, Samali A, Gorman AM (2016) The integrated stress response. EMBO Rep 17:1374–1395. https://doi.org/10.15252/embr.201642195

Pasini S, Corona C, Greene LA, Shelanski M (2015) Specific downregulation of hippocampal ATF4 reveals a necessary role in synaptic plasticity and memory. Cell Rep 11:183–191. https://doi.org/10.1016/j.celrep.2015.03.025

Prakash S, Matouschek A (2004) Protein unfolding in the cell. Trends Biochem Sci 29(11):593–600. https://doi.org/10.1016/j.tibs.2004.09.011

Rattan SIS (2008) Hormesis in aging. Ageing Res Rev 7:63–78. https://doi.org/10.1016/j.arr.2007.03.002

Reddi AR, Culotta VC (2013) SOD1 integrates signals from oxygen and glucose to repress respiration. Cell 152:224–235. https://doi.org/10.1016/j.cell.2012.11.046

Ren B, Robert F, Wyrick JJ, Aparicio O, Jennings EG, Simon I, Zeitlinger J, Schreiber J, Hannett N, Kanin E, Volkert TL, Wilson CJ, Bell SP, Young RA (2000) Genome-wide location and function of DNA binding proteins. Science 290:2306–2309. https://doi.org/10.1126/science.290.5500.2306

Rinnerthaler M, Büttner S, Laun P, Heeren G, Felder TK, Klinger H, Weinberger M, Stolze K, Grousl T, Hasek J, Benada O, Frydlova I, Klocker A, Simon-Nobbe B, Jansko B, Breitenbach-Koller H, Eisenberg T, Gourlay CW, Madeo F, Burhans WC, Breitenbach M (2012) Yno1p/Aim14p, a NADPH-oxidase ortholog, controls extramitochondrial reactive oxygen species generation, apoptosis, and actin cable formation in yeast. PNAS 109(22):8658–8663. https://doi.org/10.1073/pnas.1201629109

Rodríguez-Ezpeleta N, Brinkmann H, Burey SC, Roure B, Burger G, Löffelhardt W, Bohnert HJ, Philippe H, Lang BF (2005) Monophyly of primary photosynthetic eukaryotes: green plants, red algae, and glaucophytes. Curr Biol 15(14):1325–1330. https://doi.org/10.1016/j.cub.2005.06.040

Salwinski L, Miller CS, Smith AJ, Pettit FK, Bowie JU, Eisenberg D (2004) The database of interacting proteins: 2004 update. Nucleic Acids Res 32:D449–D451. https://doi.org/10.1093/nar/gkh086

Sánchez-Martín P, Komatsu M (2019) Physiological stress response by selective autophagy. J Mol Biol 432(1):53–62. https://doi.org/10.1016/j.jmb.2019.06.013

Schaur R, Siems W, Bresgen N, Eckl PM (2015) 4-Hydroxy-nonenal—A bioactive lipid peroxidation product. Biomolecules 5:2247–2337. https://doi.org/10.3390/biom5042247

Schrödinger E (1944) What is life? The physical aspect of the living cell. Cambridge University Press, Cambridge

Siauciunaite R, Foulkes NS, Calabrò V, Vallone D (2019) Evolution shapes the gene expression response to oxidative stress. Int J Mol Sci 20(12):3040. https://doi.org/10.3390/ijms20123040

Sopko R, Huang D, Preston N, Chua G, Papp B, Kafadar K, Snyder M, Oliver SG, Cyert M, Hughes TR, Boone C, Andrews B (2006) Mapping pathways and phenotypes by systematic gene overexpression. Mol Cell 21:319–330. https://doi.org/10.1016/j.molcel.2005.12.011

Stincone A, Prigione A, Cramer T, Wamelink MMC, Campbell K, Cheung E, Olin-Sandoval V, Grüning N-M, Krüger A, Tauqeer Alam M, Keller MA, Breitenbach M, Brindle KM, Rabinowitz JD, Ralser M (2015) The return of metabolism: biochemistry and physiology of the pentose phosphate pathway. Biol Rev 90:927–963. https://doi.org/10.1111/brv.12140

Szabo S, Tache Y, Somogyi A (2012) The legacy of Hans Selye and the origins of stress research: a retrospective 75 years after his landmark brief "Letter" to the Editor of Nature. Stress 15(5):472–478. https://doi.org/10.3109/10253890.2012.710919

Taymaz-Nikerel H, Cankorur-Cetinkaya A, Kirdar B (2016) Genome-wide transcriptional response of Saccharomyces cerevisiae to stress-induced perturbations. Front Bioeng Biotechnol 4:17. https://doi.org/10.3389/fbioe.2016.00017

Temple MD, Perrone G, Dawes IW (2005) Complex cellular responses to reactive oxygen species. Trends Cell Biol 15(6):319–326. https://doi.org/10.1016/j.tcb.2005.04.003

Teng X, Dayhoff-Brannigan M, Cheng W-C, Gilbert CE, Sing CN, Diny NL, Wheelan SJ, Dunham MJ, Boeke JD, Pineda FJ, Hardwick JM (2013) Genome-wide consequences of deleting any single gene. Mol Cell 52:485–494. https://doi.org/10.1016/j.molcel.2013.09.026

Thomas DC (2017) The phagocyte respiratory burst: historical perspectives and recent advances. Immunol Lett 192:88–96. https://doi.org/10.1016/j.imlet.2017.08.016

Thorpe GW, Fong CS, Alic N, Higgins VJ, Dawes IW (2004) Cells have distinct mechanisms to maintain protection against different reactive oxygen species: oxidative-stress-response genes. Proc Natl Acad Sci USA 101(17):6564–6569. https://doi.org/10.1073/pnas.0305888101

Tong AHY, Lesage G, Bader GD, Ding H, Xu H, Xin X, Young J, Berriz GF, Brost RL, Chang M, Chen Y, Cheng X, Chua G, Friesen H, Goldberg D, Haynes J, Humphries C, He G, Hussein S, Ke L, Krogan N, Li Z, Levinson JN, Lu H,

Ménard P, Munyana C, Parsons AB, Ryan O, Tonikian R, Roberts T, Sdicu A-M, Shapiro J, Sheikh B, Suter B, Wong SL, Zhang LV, Zhu H, Burd CG, Munro S, Sander C, Rine J, Greenblatt J, Peter M, Bretscher A, Bell G, Roth FP, Brown GW, Andrews B, Bussey H, Boone C (2004) Global mapping of the yeast genetic interaction network. Science 303:808. https://doi.org/10.1126/science.1091317

Toone WM, Jones N (1999) AP-1 transcription factors in yeast. Curr Opin Genet Dev 9:55–61. https://doi.org/10.1016/s0959-437x(99)80008-2

Tsai M-C, Huang T-L (2015) Thiobarbituric acid reactive substances (TBARS) is a state biomarker of oxidative stress in bipolar patients in a manic phase. J Affect Disord 173:22–26. https://doi.org/10.1016/j.jad.2014.10.045

Verrastro I, Pasha S, Tveen Jensen K, Pitt AR, Spickett C (2015) Mass spectrometry-based methods for identifying oxidized proteins in disease: advances and challenges. Biomolecules 5:378–411. https://doi.org/10.3390/biom5020378

Wächtershäuser G (1990) Evolution of the first metabolic cycles. Proc Natl Acad Sci USA 87(1):200–204. https://doi.org/10.1073/pnas.87.1.200

Wamelink MMC, Struys EA, Jakobs C (2008) The biochemistry, metabolism and inherited defects of the pentose phosphate pathway: a review. Metab Diss 31:703–717. https://doi.org/10.1007/s10545-008-1015-6

Winzeler EA, Shoemaker DD, Astromoff A, Liang H, Anderson K, Andre B, Bangham R, Benito R, Boeke JD, Bussey H, Chu AM, Connelly C, Davis K, Dietrich F, Dow SW, El Bakkoury M, Foury F, Friend SH, Gentalen E, Giaever G, Hegemann JH, Jones T, Laub M, Liao H, Liebundguth N, Lockhart DJ, Lucau-Danila A, Lussier M, M'Rabet N, Menard P, Mittmann M, Pai C, Rebischung C, Revuelta JL, Riles L, Roberts CJ, Ross-MaxDonals P, Scherens B, Snydier M, Sookhai-Mahadeo S, Storms RK, Vérronneau S, Voet M, Volckaert G, Ward TR, Wysocki R, Yen GS, Yu K, Zimmermann K, Philippsen P, Johnston M, Davis RW (1999) Functional characterization of the S. cerevisiae genome by gene deletion and parallel analysis. Science 285:901. https://doi.org/10.1126/science.285.5429.901

Yamamoto A, Ueda J, Yamamoto N, Hashikawa N, Sakurai H (2007) Role of heat shock transcription factor in *Saccharomyces cerevisiae* oxidative stress response. Eukaryot Cell 6(8):1373–1379. https://doi.org/10.1128/EC.00098-07

Yang W, Hekimi S (2010) A mitochondrial superoxide signal triggers increased longevity in *Caenorhabditis elegans*. PLoS Biol 8(12):1–14. https://doi.org/10.1371/journal.pbio.1000556

Yang K, Wang C, Sun T (2019) The roles of intracellular chaperone proteins, sigma receptors, in Parkinson's disease (PD) and major depressive disorder (MDD). Front Pharmacol 10:528. https://doi.org/10.3389/fphar.2019.00528

Yokoyama H, Mizunuma M, Okamoto M, Yamamoto J, Hirata D, Miyakawa T (2006) Involvement of calcineurin-dependent degradation of Yap1p in Ca2+-induced G2 cell-cycle regulation in Saccharomyces cerevisiae. EMBO Rep 7(5):519–524

Zhu PJ, Khatiwada S, Reineke LC, Dooling SW, Kim JJ, Lei W, Walter P, Costa-Mattioli M (2019) Activation of the ISR mediates the behavioral and neurophysiological abnormalities in Down syndrome. Science 366:843–849. https://doi.org/10.1126/science.aaw5185

Zito E (2015) ERO1: a protein disulfide oxidase and H2O2 producer. Free Radic Biol Med 83:299–304. https://doi.org/10.1016/j.freeradbiomed.2015.01.011

4

Oxidative Stress and the Brain: A Working Hypothesis for the Generation of Psychological Stress

While we have seen in Chap. 3 that stress is an ambivalent phenomenon that is at the same time unavoidable and necessary for life and—if exceeding an organism's capacities to respond to stress—life threatening. In this chapter, we turn to the often-destructive effects and detrimental consequences of chronic stress. Focusing on different regions of the brain, we will describe manifestations of chronic stress in terms of "typical" stress-related syndromes and diseases, such as **major depression**,[1] psychosomatic heart diseases, and the impact of stress on the immune system. The findings from research we draw on in this chapter concerning such adverse outcomes of stress exposure clearly touch not solely on the physiology of stress but also on stress's psychological side. This is a crucial aspect when we address the stress of poverty. Before going into a deeper conversation between stress research and poverty studies in later chapters of this book, here we will present some anchor points on the relationship of chronic stress and the experience of poverty, and provide a working hypothesis for the generation of psychological stress.

Redox Metabolism in the Brain

As we have seen, from a biochemical perspective the best-studied signaling networks mediating the **stress response** are the integrated stress response (ISR) on the one hand, and the hydrogen peroxide (H_2O_2) mediated

---

[1] Glossary terms are bolded at first mention in each chapter.

© Springer Nature Switzerland AG 2021
M. Breitenbach et al., *Stress and Poverty*, https://doi.org/10.1007/978-3-030-77738-8_4

signaling in response to the **epidermal growth factor** and other growth factors, on the other.

The integrated stress response occurs in a large variety of cellular stress situations, working through a central hub based on phosphorylation of the **translation** factor **eIF2alpha** and leading to transcriptional activation of a whole battery of stress response genes (Pakos-Zebrucka et al. 2016). The response to the epidermal growth factor is modified in relation to cellular stress based on hydrogen peroxide reversibly oxidizing cysteine sulfhydryl (SH) groups on a phosphoprotein phosphatase, as summarized in the preceding chapter.

Although the role of **ROS** and of H_2O_2 in the brain and their possible role in neuronal signal transduction is not known in molecular detail, research is now progressing to being able to study the role of **oxidative stress** in the brain while under psychological stress, which is the main prerequisite for understanding the biological basis of the pathology and sociology of psychological stress.

The following is a working model for the explanation of the destructive activity of prolonged psychological stress in terms of what we know about the biochemical nature of cellular stress and the current understanding of neuronal signal transduction in the central nervous system.

The human brain is exceptional because of a peculiar metabolic division of labor between neural and **glia cells** (**astrocytes**) (Allaman et al. 2011). Very briefly, the demand for biological energy (**ATP**) in nerve cells is very high—comparable to and exceeding the heart muscle, per gram of tissue. For the brain, glucose is practically the only carbon and energy source. In the astrocytes, which are in intimate contact with neurons, glycolysis is conducted to the point of **lactate**, which is then secreted and taken up by neurons, converted back to pyruvate, metabolized by pyruvate dehydrogenase and later by **mitochondrial** respiration, producing ATP with much higher efficiency than by glycolysis alone. This pathway of ATP production works in the waking brain at full efficiency, also using, of course, a stoichiometric amount of oxygen. Under these conditions, potentially dangerous amounts of ROS (initially superoxide which is converted to H_2O_2) are produced as a side reaction of cellular respiration. Oxygen and ROS play a very special role in the brain as compared to all other tissues of the body. The brain is dependent on oxygen, however. Even a short deficiency of oxygen, in the order of minutes, can lead to cell death in the brain, as is shown, for instance, in the aftermath of a stroke (Dowling 2018). But as well, the neurons of the brain are particularly sensitive to oxygen and to ROS (Angelova and Abramov 2018; Cobley et al. 2018). To give an example, one consequence of this is that the maternally inherited "**mitochondrial**" **diseases** caused by mutations in mitochondrial DNA lead to severe neurological symptoms. In one case (Leber hereditary optic neuropathy, LHON) the mitochondrial defect leads to blindness due to ATP scarcity in the optic nerve.

To counteract the oxidative stress produced by neuronal metabolism, glycolysis in neuronal cells is diverted from its main branch by inactivating **phosphoglucose isomerase** allosterically. This results in diverting the flow of metabolites to **glucose-6-phosphate dehydrogenase** and production of **NADPH** in the pentose phosphate cycle, which is needed to maintain the redox equilibrium in the cells. The pathways of the antioxidative response (see Chap. 3) are active in neurons, including biosynthesis and redox regulation of **glutathione**, which is produced at a high level in major regions of the brain (Rae and Williams 2017).

The Basics of Stress in Neurobiology

Psychological stress is a condition of the conscious mind and can only be built up via impressions of the senses (sensory input), memory, and information storage in the brain. The **hippocampus** is the central region of the brain where memory is created and short-term memory is stored. However, long-term memory cannot be easily located; it is not stored in the hippocampus, but probably needs a dynamic and combinatorial cooperation between a number of different brain regions. These textbook facts of neurobiology are backed up by a large number of clinical observations and experiments performed with rodents in the laboratory (Dowling 2018). During early postnatal development, but also in the course of the formation of long-term memory, the neuronal cells of the hippocampus form new synapses. In adult life, **long-term potentiation** (**LTP**), which is necessary for learning, requires "**neuronal plasticity**." This technical term refers to the ability to facilitate signal transmission through repeated signaling at a synapse and to form new synapses, even in the adult brain (Dowling 2018).

Long-term potentiation is apparently needed to form memory (although definite proof is still missing). This process results in an increase in the readiness to depolarize and in the amount and duration of **depolarization** (measured in mV) in the postsynaptic neuron by previous synaptic signaling. The experimental findings and the nomenclature created to describe these phenomena are unfortunately somewhat confusing. One LTP is nevertheless "short-term" or "early" and is attained without de novo protein synthesis. The memory of this kind lasts only for seconds. There is also a "long-term" ("late") LTP, which does need de novo protein synthesis (experimentally tested with protein synthesis inhibitors in rodents), and leads to memory lasting for minutes and hours (Dowling 2018). Finally, the nervous activity of the hippocampus, which we have just described, leads to memory storage that is not confined to the hippocampus but needs the above-mentioned LTP mechanism to create a permanent memory. We do not know, however, exactly where it is located or what the exact physical substrate of this very long-term memory is.

Also located in the hippocampus, in addition to LTP, is its counteracting mechanism, **long-term depression** (**LTD**); not to be confused with depression as a disease, it is a prolonged decrease in the amount and duration of synaptic depolarization. If and when the compensatory LTD can take place depends on the identity of the presynaptic cell and on its previous history. LTD in the hippocampus is possibly caused by the inhibitory neurotransmitter **gamma-amino butyric acid** (**GABA**). Both LTP and LTD are necessary to create new memories.

The hippocampus is the part of the brain where brain stem cells, de novo formation and regeneration of brain tissue have been observed. This finding was controversial in the literature for about 20 years but is now generally accepted (Gage 2019). What is still controversial, however, is whether or not this process is needed for memory formation.

A Biochemical Working Hypothesis for Psychological Stress

What follows is a sequence of events that describe our working hypothesis to explain how psychological stress is created.

A stressful situation starts from very large (traumatic) or repeated sensory input (chronic stress). In neurophysiological terms, this means that the repeated firing of a synapse using the excitatory neurotransmitter, glutamate, leads to depolarization of the postsynaptic cell. Entry of Na^+ through ion channels is coupled to the two voltage-dependent glutamate receptor types (mentioned below) on the postsynaptic side of the synaptic cleft. The first of these two receptors/ion channels is specific for Na^+ and can be stimulated by **AMPA**, but not by **NMDA**. The second glutamate receptor can be stimulated by NMDA, but not by AMPA, and allows the inflow of not only Na^+ but also Ca^{++} (Kandel and Mack 2013). The increased firing frequency of the synapse leads to an increased demand for ATP. Eventually, through the glutamate receptor **NMDA channel**, calcium ions enter the cell, a process which is necessary for LTP, but can also cause oxidative stress, depending on the dose and timing (Orrenius et al. 2003; Choi 1995). The source of the oxidative stress created is very likely the large increase in respiratory chain activity under these conditions. Other sources of ROS in the active neurons, such as **NADPH oxidases** and monoamine oxidases should also be considered, but have not been studied in detail in connection with psychological stress.

The effect of the above-described hyperactivity situation is a massive loss of neuronal plasticity and in the extreme, pathological, case the outcome is cell death through **apoptosis**. This is generally called neurodegeneration: loss of neuroplasticity and programmed cell death leading to the well-described defects in cognition, learning, and emotion caused by chronic psychological stress.

Oxidative stress through increased, repeated, chronic excitatory stimulation of glutamate receptors in postsynaptic cells in the hippocampus is the first step of a process leading to calcium ion inflow, increased respiratory activity, and ROS generation.

However, this process is not isolated, but embedded in a larger stress response system. This includes the integrated stress response system (ISR) with its central hub, the eIF2alpha trigger, and the downstream Atf4 transcription factor. Moreover, it is embedded in the hormonal stress system of the **HPA axis** and the sympathetic nervous system leading to **epinephrine** secretion (see Fig. 1.1). Importantly, the brain cells involved are responsive to stress hormones, e.g., to **cortisol** and to the **corticotropin-releasing hormone** (**CRH**) (see next paragraph).

Further Components of the Psychological Stress Response

In elaborate animal experiments, such as the Morris water maze for memory-training mice, it has been shown that previously experienced stress prevents or weakens cognitive function and memory in the animals. Two common methods for exposing the mice to social stress are adding an unrelated mouse to a group of mice or exposing mice to a new dominant individual. In the water maze, the previously trained mouse is tested to see if, after having been exposed to stress, it can remember where the underwater platform is, and how quickly it can reach this platform. Memory training under stress generally, and in this experimental setup, in particular, impedes or slows the formation of memory.

Little is known in detail about how the neurological and psychological stress described up to now influences an individual's **endocrine system**. We think it is very likely that the main pathway of interaction between the two systems (endocrine and neurological) is the **autonomic nervous system**, composed of the sympathetic and **parasympathetic nervous systems**. Very generally, these two antagonistic systems, which are to a large degree "involuntary," come into play. The sympathetic nervous system activates the "fight or flight" stress response via the **thalamus, amygdala**, and **hypothalamus**. However, the central **agency** governing the involuntary sympathetic nervous system is the noradrenergic **locus coeruleus** of the brain stem (Samuels and Szabadi 2008a, b), which among many other targets aims to reach the **adrenal medulla**, producing the stress hormone, epinephrine. Many of the complex stress reactions activated are under the direct control of the sympathetic nervous system, but many are activated by the HPA axis starting in the hypothalamus, under hormonal control through cortisol. Conversely, cortisol exerts a feedback reaction through cortisol receptors in the brain, and in particular in the hippocampus. Also, the main releasing factor of the HPA axis, CRH, has receptors in many parts of the brain (Tsigos and Chrousos 2002). This could be part of the feedback dampening of an excessive stress reaction.

On the other hand, roughly speaking, the parasympathetic nervous system is doing the opposite of the sympathetic system, governing the "rest and digest" reaction. The other branch of the stress reaction, in addition to the HPA axis and cortisol, involves epinephrine, a hormone mainly responsible for activating metabolic reserves, like liver glycogen, and produced under sympathetic neuronal control by the adrenal medulla. Cortisol and epinephrine together are responsible for the full-blown "fight or flight" response (compare Fig. 1.1 in Chap. 1). Like the cortisol and CRH receptors, epinephrine receptors are common in the brain and in the nervous system in general. The sympathetic autonomic nervous system produces **norepinephrine**, a neurohormone that acts on the target cells (in part these are the same cells as for epinephrine) in two independent ways, namely directly via the nerve endings targeting the organs, and also as an endocrine hormone via the blood flow. Activation of the system as a consequence of sensory input is accomplished via a chain: thalamus, amygdala, hypothalamus, *locus coeruleus*, and the autonomic nervous system.

As a whole, the neuroendocrine system described so far is regulated by a still not completely understood network of feedback and feed-forward reactions, which in normal conditions serves to maintain **homeostasis** and to allow often life-saving stress reactions as well as, eventually, complete recovery from stress. This network encompasses the autonomic nervous system and the neuroendocrine system, including the hormones of the **adrenal cortex** and medulla.

Examples of the Role of Psychological Stress in Psychosomatic Diseases

We hypothesize that chronic stress leads to a condition where recovery is no longer possible (at least not in the short term), leading to the diseases of stress, which the seminal papers of Selye mentioned and a large number of medical investigations have elucidated in recent decades.

To bring diseases of chronic stress into focus, we will discuss some typical psychosomatic pathologies for which the causative action of stress is generally accepted and well researched, namely: major depression, psychosomatic heart disease, and the impact of stress on the immune system. The interrelation of psychosomatic pathologies with poverty has already been noted (Baum et al. 1999; Sundquist et al. 2004) and will be further discussed with more descriptive examples in later chapters of this book.

Major Depression

First, we will discuss the etiology of major depression (Yang et al. 2015; Lee et al. 2002).

Among the most prevalent psychiatric diseases recognized today, major depression has a relatively small hereditary contribution of 40% or less, as determined by careful twin studies (Kandel 2018). This means that the environmental, lifestyle, and life-history causes of the disease are more important than a possible genetic predisposition (Border et al. 2019). The neuroendocrinological status of a patient in major depression is very similar to a person's chronic stress status (Holsboer 2000; Pariante and Lightman 2008). This includes the activity of the HPA axis and cortisol levels (see Chap. 6). Some typical causative factors in the life history of patients are the loss of a job or the loss of loved ones, and as we argue here, life in chronic poverty, which produces numerous secondary factors that can also cause stress and depression, such as detrimental environmental conditions, disrupted family life, and substance abuse. These examples only hint at a much richer story of life-history events and their interactions, which can cause major depression in poverty. This will be described and analyzed in much more detail in the chapters dealing explicitly with the psychology and sociology of poverty.

Major depression is amenable to pharmacological treatment. We will mention only one of several pharmacological possibilities, namely the antidepressant drugs that prevent serotonin reuptake in synapses of the hippocampus. The high cortisol level found in depressive patients can inhibit tryptophan degradation and consequently serotonin synthesis. Disruption of neuronal signaling through glutamate (and its antagonist GABA) apparently also can disturb the concentration of the neurotransmitter serotonin, which becomes too low. This can be counteracted by prohibiting serotonin reuptake into the presynaptic cells, targeting the serotonin transporter SLC6A4 with the specific serotonin reuptake drugs (**SSRI**) (Zhu et al. 2017). This form of pharmacological treatment is successful in an appreciable percentage of depressive patients. Why not in 100% of those patients? Presumably, the neurochemical causes of depression are not completely uniform in all individuals. It is an open question at the moment whether all depressive patients who present with symptoms of chronic stress respond to this form of therapy. The side effects of the treatment of major depression with serotonin reuptake inhibiting drugs are not negligible and are discussed in the papers cited (for instance, Zhu et al. 2017). Needless to say, pharmacological treatment of depression is not a preferred method to fight poverty. From the perspective of poverty studies, the treatment, however successful, of major depression as a potential consequence of the experience of chronic poverty stress (see later chapters of the book) must not be mistaken for the "treatment" or successful combating of poverty itself.

Psychosomatic Heart Disease

Second, we will discuss how psychological stress causes and aggravates certain types of heart disease.

In popular belief, **coronary heart disease** and death resulting from heart infarction and stroke are classical diseases strongly influenced by psychosocial factors. Results of the INTERHEART study dealing with this question confirmed and quantified this phenomenon (Yusuf et al. 2004; Rosengren et al. 2004; Lagraauw et al. 2015). Coronary heart disease has a strong and indisputable etiology in lifestyle (diet, physical activity, smoking, and drinking) with a known causality based on **atherosclerosis** (not discussed in detail here), but in addition, it is very highly correlated with environmental and endogenous stress factors. A schematic picturing this interrelationship is shown in Fig. 4.1. This figure visualizes the complex relationship between three causative agents that can influence each other reciprocally in many typical patients suffering from heart disease: lifestyle factors, psychological stress, and chronic ischemic heart disease resulting from the first two factors. The lifestyle factors include smoking, drinking, **body mass index** (**BMI**), sedentary lifestyle, and lack of balanced and adequate nutrition. The psychological stress factors are multiple; one typical example would be nighttime noise (Münzel et al. 2014). In chronic ischemic heart disease, which can be caused by either of the other two factors, or by a combination of both, we see atherosclerosis, the narrowing of and in an extreme case obstructing coronary arteries. In addition, there may be a genetic component in chronic heart disease that we will not discuss here. Psychological or emotional stress can often trigger acute cardiac failure and even death in a patient predisposed by the causative factors described so far (Steptoe and Brydon 2009). There is strong

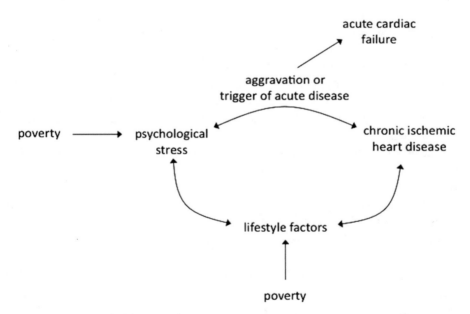

Fig. 4.1 The complex interactions relevant to heart disease. As shown here, poverty can produce psychological stress, and it also has a strong influence on lifestyle factors, such as smoking, drinking, unhealthy diet, and sedentary behavior. Lifestyle factors and psychological stress mutually reinforce each other and can induce chronic ischemic heart disease. Typically leading to reduced physical fitness, this condition can produce psychological stress (hence the mutualistic interaction with psychological stress). This chronic condition can turn into acute and life-threatening heart failure, often triggered by psychological stress or major catastrophic events experienced by the individual

evidence for the influence of poverty on psychological stress as well as on lifestyle factors. There are, however, acute forms of heart disease that can be triggered by psychological stress even in the absence of chronic ischemic coronary heart disease; an example is the **Takotsubo syndrome**, discussed later in this chapter.

In the INTERHEART study, more than 11,000 patients who had had a first myocardial infarction and 13,000 age- and sex-matched controls from 52 different countries were systematically clinically investigated and their stress factors evaluated by questionnaire. The stress factors found to be significantly correlated with coronary heart disease and heart infarction were independent of ethnicity, general cultural factors, and the sex of the patient. A significant correlation was found with major depression (discussed above). The main questions that were asked to determine the stress level of the patients focused on the year preceding the infarction. These questions and stress factors are important for our purposes, namely: financial stress (such as the loss of a business or business failure), workplace stress, retirement, violence, familial stress (major interfamily conflict, marital separation, divorce, death of partner, death or major illness of a close relative), and major accidents or injuries. The socioeconomic status of the patients and controls was asked for but the data were not analyzed in detail to determine the significance of the relation between socioeconomic status (SES) and cardiovascular disease.

Two additional relevant observations are described in the current cardiovascular medical literature.

A clearly defined heart disease is stress-induced cardiomyopathy, a specific deformation of (in most cases) the left ventricle caused by severe and acute stress resulting in a catecholamine surge. In the literature, this syndrome is also often referred to as "apical ballooning syndrome" or "Takotsubo syndrome," named after the form of patients' dilated hearts (Fig. 4.2), which resemble traditional Japanese octopus traps (Akashi et al. 2015; Keramida et al. 2020; Schneider et al. 2014). Its exact pathogenesis and pathophysiology are still incompletely understood (Lyon 2017; Möhlenkamp et al. 2020). However, its stress-related etiology is particularly well documented, both concerning psychological and physiological triggers (the latter are also documented in animal experiments). A popular and telling name used for the Takotsubo syndrome is "broken heart syndrome"; this is due to the fact that it is often triggered by severe emotional stress—by bereavement, but also by **stressors** as mentioned above with regard to stress-related coronary heart diseases, such as job loss or severe financial strain (Akashi et al. 2015; Möhlenkamp et al. 2020). In some cases, there is an overlap with coronary heart disease, but

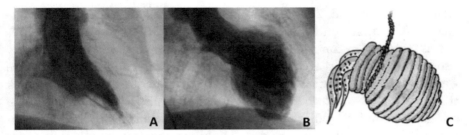

Fig. 4.2 The Takotsubo cardiomyopathy. Part **a** shows an X-ray of a normal heart. In part **b**, the X-ray shows that the left ventricle of the Takotsubo patient's heart is enlarged, leading to a diminished pumping function. Part **c** shows a "takotsubo," a traditional Japanese octopus trap. Because the dilated heart is similar in shape, the syndrome is often called "Takotsubo cardiomyopathy." Source: Schneider (2012) with permission of contributors: © B. Schneider for **a** and **b**, © J. Neuffer for **c**

there are many cases where the pathology of the coronary arteries is absent, and complete recovery after this life-threatening disease is possible.

In recently described statistical investigations it has been found that major catastrophic events can induce an increase in the number of the aforementioned heart diseases in the general population, even if the patients had no direct contact with the catastrophe, but just learned of it from the media. A highly significant increase in heart infarctions occurred in the days after the 9/11 attacks on the New York World Trade Center (Allegra et al. 2005) and in the days after rocket attacks on Tel Aviv during the Gulf War (Rivkind et al. 1997), for instance. In the case of stress-induced Takotsubo disease, there is also evidence of "small epidemics" following natural disasters such as major earthquakes or hurricanes (Akashi et al. 2015).

Stress and the Immune System

Third, we will discuss existing evidence for interaction between stress and the immune system (Dhabhar 2014; Pierre 2019). Exposed to (chronic) psychological stress, the brain can interfere with the immune system and inhibit its functions (Straub and Cutolo 2018). Experimental evidence shows that acute stress can repurpose the **innate immune** system by redirecting the cellular immune response (monocytes, NK cells, etc.) to the location where it is most needed, e.g., the skin (Dhabhar 2014). However, chronic stress leads to a general deficit in both the innate and adaptive immune system, making patients, including cancer patients (Sephton et al. 2009), more susceptible to infection (Smith et al. 2004; Vissoci Reiche et al. 2004; Glaser and Kiecolt-Glaser 2005; Straub and Cutolo 2018; Seiler et al. 2020).

It is tempting to conclude that in this respect the chronic stress of poverty also leads to a health risk concerning infectious diseases and cancer in people with low socioeconomic status. In fact, in a broad literature review, Elwenspoek et al. (2017) have shown how exposure to early-life adversities, including being raised under conditions of poverty (in households with low socioeconomic status), has tremendous effects on the immune system. They also refer to adult health risks, including psychiatric and cardiovascular disorders as well as chronic inflammatory disorders, such as asthma and allergies, or autoimmune disorders (see also Danese and McEwen 2012; Fagundes and Way 2014). Chronic early-life stress may lead to epigenetic changes that last over the lifespan (and potentially over generations) and still affect immune reactivity in adult life (Fagundes and Way 2014). Adverse childhood experiences as well as low SES as adults are also found to have an impact on certain types of cancer (Steel et al. 2020) and on mortality in various cancers (Gilligan 2005).

Conclusion

As we conclude this chapter describing our unifying working hypothesis concerning the generation of psychological stress and the effects of psychological stress on health, we consider the relevance of these neurophysiological investigations for the psychology and sociology of the stress of poverty. Generally speaking, published medical and biochemical articles dealing directly and explicitly with this correlation are scarce, but psychological and sociological articles are abundant (see later chapters). The correlation of stress with a growing number and increased severity of diseases as well as the correlation of poverty with chronic psychological stress is highly significant.

What can be learned from all these facts for our topic of "stress and poverty"? First of all, poverty is for most people chronic stress. Importantly, this is even more true for children. How does this impact their later life? **Epigenetics** (discussed in the next chapter) can create a kind of predisposition for certain diseases. Typical examples are **metabolic syndrome** and type II diabetes, but also heart disease. It is not easy under conditions of poverty to realize one's potential to the full—intellectually and in all other parts of life. In this way, through a mechanism often called "cultural inheritance," the adverse experience of poverty and deprivation is perpetuated for many people.

References

Akashi Y, Nef HM, Lyon AR (2015) Epidemiology and pathophysiology of Takotsubo syndrome. Nat Rev Cardiol 12(7):387–397. https://doi.org/10.1038/nrcardio.2015.39

Allaman I, Bélanger M, Magistretti PJ (2011) Astrocyte–neuron metabolic relationships: for better and for worse. Trends Neurosci 34(2):76–87. https://doi.org/10.1016/j.tins.2010.12.001

Allegra JR, Mostashari F, Rothman J, Milano P, Cochrane DG (2005) Cardiac events in New Jersey after the September 11, 2001, terrorist attack. J Urban Health 82(3):358–363. https://doi.org/10.1093/jurban/jti087

Angelova PR, Abramov AY (2018) Role of mitochondrial ROS in the brain: from physiology to neurodegeneration. FEBS Lett 592:692–702. https://doi.org/10.1002/1873-3468.12964

Baum A, Garofalo JP, Yali AM (1999) Socioeconomic status and chronic stress: does stress account for SES effects on health? Ann N Y Acad Sci 896:131–144

Border R, Johnson EC, Evans LM, Smolen A, Berley N, Sullivan PF, Keller MC (2019) No support for historical candidate gene or candidate gene-by-interaction hypotheses for major depression across multiple large samples. Am J Psychiatry 176(5):376–387. https://doi.org/10.1176/appi.ajp.2018.18070881

Choi DW (1995) Calcium: still center-stage in hypoxic-ischemic neuronal death. Trends Neurosci 18(2):59–60

Cobley JN, Fiorello ML, Bailey DM (2018) 13 reasons why the brain is susceptible to oxidative stress. Redox Biol 15:490–503. https://doi.org/10.1016/j.redox.2018.01.008

Danese A, McEwen BS (2012) Adverse childhood experiences, allostasis, allostatic load, and age-related disease. Physiol Behav 106:29–39. https://doi.org/10.1016/j.physbeh.2011.08.019

Dhabhar FS (2014) Effects of stress on immune function: the good, the bad, and the beautiful. Stanford Immunol 58:193–210. https://doi.org/10.1007/s12026-014-8517-0

Dowling JE (2018) Understanding the brain: from cells to behavior to cognition. Norton, New York

Elwenspoek MMC, Kuehn A, Muller CP, Turner JD (2017) The effects of early life adversity on the immune system. Psychoneuroendocrinology 82:140–154. https://doi.org/10.1016/j.psyneuen.2017.05.012

Fagundes CP, Way B (2014) Early-life stress and adult inflammation. Curr Dir Psychol Sci 23(4):277–283. https://doi.org/10.1177/0963721414535603

Gage FH (2019) Adult neurogenesis in mammals: neurogenesis in adulthood has implications for sense of self, memory, and disease. Science 364(6443):827–828. https://doi.org/10.1126/science.aav6885

Gilligan T (2005) Social disparities and prostate cancer: mapping the gaps in our knowledge. Cancer Causes Control 16:45–53

Glaser R, Kiecolt-Glaser JK (2005) Stress-induced immune dysfunction: implications for health. Nat Rev Immunol 5:243–251. https://doi.org/10.1038/nri1571

Holsboer F (2000) The corticosteroid receptor hypothesis of depression. Neuropsychopharmacology 23(5):477–501

Kandel ER (2018) The disordered mind: what unusual brains tell us about ourselves. Robinson, London

Kandel ER, Mack S (2013) Principles of neural science. McGraw-Hill Medical, New York

Keramida K, Backs J, Bossone E, Citro R, Dawson D, Omerovic E, Parodi G, Schneider B, Ghadri JR, Van Laake LW, Lyon AR (2020) Takotsubo syndrome in heart failure and world congress on acute heart failure 2019: highlights from the experts. ESC Heart Fail 7:400–406. https://doi.org/10.1002/ehf2.12603

Lagraauw HM, Kuiper J, Bot I (2015) Acute and chronic psychological stress as risk factors for cardiovascular disease: insights gained from epidemiological, clinical and experimental studies. Brain Behav Immun 50:18–30. https://doi.org/10.1016/j.bbi.2015.08.007

Lee AL, Ogle WO, Sapolsky RM (2002) Stress and depression: possible links to neuron death in the hippocampus. Bipolar Disord 4:117–128

Lyon A (2017) Stress in a dish. Exploring the mechanisms of Takotsubo syndrome. J Am Coll Cardiol 70(8):992–995

Möhlenkamp S, Kleinbongard P, Erbel R (2020) Tako-Tsubo-Syndrom. Der Kardiologe 14:323–336. https://doi.org/10.1007/s12181-020-00415-y

Münzel T, Gori T, Babisch W, Basner M (2014) Cardiovascular effects of environmental noise exposure. Eur Heart J 35:829–836. https://doi.org/10.1093/eurheartj/ehu030

Orrenius S, Zhivotovsky B, Nicotera P (2003) Regulation of cell death: the calcium-apoptosis link. Nat Rev Mol Cell Biol 4:552–565. https://doi.org/10.1038/nrm1150

Pakos-Zebrucka K, Koryga I, Mnich K, Ljujic M, Samali A, Gorman AM (2016) The integrated stress response. EMBO Rep 17(10):1374–1395. https://doi.org/10.15252/embr.201642195

Pariante CM, Lightman SL (2008) The HPA axis in major depression: classical theories and new developments. Trends Neurosci 31(9):464–468. https://doi.org/10.1016/j.tins.2008.06.006

Pierre P (2019) Integrating stress responses and immunity: stress is required to assemble immune signaling complexes during infection. Science 365(6448):28–29. https://doi.org/10.1126/science.aay0987

Rae CD, Williams SR (2017) Glutathione in the human brain: review of its roles and measurement by magnetic resonance spectroscopy. Anal Biochem 529:127–143. https://doi.org/10.1016/j.ab.2016.12.022

Rivkind A, Barach P, Israeli A, Berdugo M, Richter ED (1997) Emergency preparedness and responses in Israel during the Gulf War. Emerg Med 30(4):513–521

Rosengren A, Hawken A, Ôunpuu S, Sliwa K, Zubaid M, Almahmeed WA, Ngu Blackett K, Sitthi-amorn C, Sato H, Yusuf S (2004) Association of psychosocial risk factors with risk of acute myocardial infarction in 11 119 cases and 13 648 controls from 52 countries (the INTERHEART study): case-control study. Lancet 364(11):953–962

Samuels ER, Szabadi E (2008a) Functional neuroanatomy of the noradrenergic locus coeruleus: its roles in the regulation of arousal and autonomic function part I: principles of functional organisation. Curr Neuropharmacol 6:235–253

Samuels ER, Szabadi E (2008b) Functional neuroanatomy of the noradrenergic locus coeruleus: its roles in the regulation of arousal and autonomic function Part II: physiological and pharmacological manipulations and pathological alterations of locus coeruleus activity in humans. Curr Neuropharmacol 6:254–285

Schneider B (2012) Wenn das Herz zerbricht. Die Tako-Tsubo-Kardiomyopathie. HERZ HEUTE. Zeitschrift der Deutschen Herzstiftung e.V. 32(2):4–8

Schneider B, Athanasiadis A, Schwab J, Pistner W, Gottwald U, Schoeller R, Toepel W, Winter K-D, Stellbrink C, Müller-Honold T, Wegner C, Sechtem U (2014) Complications in the clinical course of tako-tsubo cardiomyopathy. Int J Cardiol 176:199–205. https://doi.org/10.1016/j.ijcard.2014.07.002

Seiler A, Fagundes CP, Christian LM (2020) The impact of everyday stressors on the immune system and health. In Choukér A (ed) Stress, challenges and immunity in space. Springer, Berlin, pp 71–92. https://doi.org/10.1007/978-3-030-16996-1_6

Sephton SE, Dhabhar FS, Keuroghlian AS, Giese-Davis J, McEwen BS, Ionan AC, Spiegel D (2009) Depression, cortisol, and suppressed cell-mediated immunity in metastatic breast cancer. Brain Behav Immun 23:1148–1155. https://doi.org/10.1016/j.bbi.2009.07.007

Smith A, Vollmer-Conna U, Bennett B, Wakefield D, Hickie I, Lloyd A (2004) The relationship between distress and the development of a primary immune response to a novel antigen. Brain Behav Immun 18:65–75. https://doi.org/10.1016/S0889-1591(03)00107-7

Steel JL, Antoni M, Pathak R, Butterfield LH, Vodovotz Y, Savkova A, Wallis M, Wang Y, Grammer E, Burke R, Brady M, Geller DA (2020) Adverse childhood experiences (ACEs), cell-mediated immunity, and survival in the context of cancer. Brain Behav Immun 88:566–572. https://doi.org/10.1016/j.bbi.2020.04.050

Steptoe A, Brydon L (2009) Emotional triggering of cardiac events. Neurosci Biobehav Rev 33:63–70

Straub RH, Cutolo M (2018) Psychoneuroimmunology—developments in stress research. Wiener Medizinische Wochenschrift 168:76–84. https://doi.org/10.1007/s10354-017-0574-2

Sundquist K, Winkleby M, Ahlén H, Johansson S-E (2004) Neighborhood socioeconomic environment and incidence of coronary heart disease: a follow-up study of 25,319 women and men in Sweden. Am J Epidemiol 159(7):655–662

Tsigos C, Chrousos GP (2002) Hypothalamic–pituitary–adrenal axis, neuroendocrine factors and stress. J Psychosom Res 53:865–871

Vissoci Reiche EM, Odebrecht Vargas Nunes S, Kaminami Morimoto H (2004) Stress, depression, the immune system, and cancer. Lancet Oncol 5:617–625

Yang L, Zhao Y, Wang Y, Liu L, Zhang X, Li B, Cui R (2015) The effects of psychological stress on depression. Curr Neuropharmacol 13(4):494–504

Yusuf S, Hawken S, Ôunpuu S, Dans T, Avezum A, Lanas F, McQueen M, Budaj A, Pais P, Varigos J, Lisheng L (2004) Effect of potentially modifiable risk factors associated with myocardial infarction in 52 countries (the INTERHEART study): case-control study. Lancet 364(11):937–952

Zhu J, Klein-Fedyshin M, Stevenson JM (2017) Serotonin transporter gene polymorphisms and selective serotonin reuptake inhibitor tolerability: review of pharmacogenetic evidence. Pharmacotherapy 37(9):1089–1104. https://doi. org/10.1002/phar.1978

5

Epigenetics and Some Further Observations on Stress-Induced Diseases

Lasting stress (physiological as well as psychological) does not affect people as "islands." As we have seen, stress and the **stress response**[1] can in numerous ways affect people beyond their own scope of action; this applies even in the sense of transgenerational transmission of pathologies resulting from chronic and overwhelming stress exposure. In this chapter, we enter a relatively young strand of science to discuss the role of **epigenetics** in stress research. In particular, we focus on detrimental health outcomes related to stress, where epigenetics may be shown to come into play, according to research findings in a variety of fields. We start this chapter by giving a clear and concise definition of epigenetics and by presenting a few well-known examples where epigenetic changes found to occur in prenatal and early postnatal life can cause a predisposition to certain diseases throughout life. In addition, our examples show that the triggers causing such epigenetic changes can be linked to the experience of poverty. This aspect will be discussed thoroughly in Chap. 8.

Definitions and Overview

Epigenetics is a term and a concept of genetic research, coined by researchers in developmental genetics, most importantly by Waddington (1942, 1957; cited in Goldberg et al. 2007). After decades of being a synonym for

---

[1] Glossary terms are bolded at first mention in each chapter.

© Springer Nature Switzerland AG 2021
M. Breitenbach et al., *Stress and Poverty*, https://doi.org/10.1007/978-3-030-77738-8_5

as-yet-unexplained facts of heredity,[2] epigenetics in recent years has become a respected topic in genetic research. The salient features necessary to define a fact of inheritance as epigenetic are the following: (1) the phenomenon is caused by a change in chromatin structure leading to a change in **gene expression** without any change in genomic DNA sequence, during embryonic and postnatal development, for instance; and (2) this change must be inherited, not only during subsequent cell division cycles, but also **intergenerationally**; i.e., it must be transmitted to the next generation. In the current literature, the second part of this strict definition is often neglected: i.e., the term "epigenetic" is often used if a change in gene expression is permanent during the life of one individual; in other words, if such a change in gene expression is relatively stable and occurs without any change in the genome sequence. The epigenetic changes are often called "**epigenetic marks**." The biochemical mechanisms leading to the inheritance of such epigenetic marks during mitotic cell cycles have been well studied, although many open questions remain (Messerschmidt et al. 2014). However, intergenerational inheritance of epigenetic marks has also been studied, as described below.

The chemical changes in chromatin structure that are most commonly encountered are **CpG methylation**, binding of regulatory RNAs, and covalent modification of histones, like lysine methylation, acetylation, deacetylation, ubiquitination, and others. CpG methylation has been shown to be the most permanent epigenetic mark that can therefore be responsible for intergenerational effects (Gluckman et al. 2007). As mentioned before, it is important to note that these changes occur without any change in the genomic DNA sequence.

In the human genome and in other mammalian genomes, the most important epigenetic mark is CpG methylation (Goll and Bestor 2005), resulting in 5-methyl cytosine on both strands of the DNA molecule. Methylation is a mark, but not the final mechanism of gene silencing, if it is present in the upstream promoter region of mammalian genes. It is well understood how this mark on DNA is transmitted through many somatic cell division cycles by means of maintenance methylation enzymes and the methyl transferring coenzyme, S-adenosyl methionine (Goll and Bestor 2005).

Around the time of nidation (implantation) of the mammalian embryo in the uterus wall, all CpG methylation marks are erased, and a new set of epigenetic marks has to be established during the subsequent embryonal and fetal life (Reik et al. 2001). Nidation is a very early process in pregnancy,

---

[2] Compare with the examples given by Goldberg et al. (2007), such as, for instance, position effect variegation in *Drosophila*.

starting around day 6 after fertilization. It is presently unknown how the information transfer that seems to be necessary for de novo methylation is achieved at this stage, re-establishing CpG methylation marks on the fetal genome. A similar critical and underexplored step for DNA methylation and demethylation is female and male **gametogenesis**. During gametogenesis, the somatic methylation pattern of the cells, which has existed since early embryogenesis, has to be erased, followed by sex-specific **imprinting** of the genome of **primordial germ cells**. What is known in detail today about these processes comes from experiments in the mouse, concerning about 70 genes. We can with good reason assume similarities in these processes so essential for life and fertility in all mammals, including human beings (Peters 2014).

The process of genome methylation is biochemically well known, and the enzymes responsible have been identified and characterized, although many open questions remain; for example, about the necessary information transfer in the de novo methylation after erasure, as mentioned above. The process of demethylation is not well known and still carries a question mark. Evidence exists that the pathway of demethylation involves a series of reactions, including the synthesis of hydroxymethyl cytosine and, consequently, removal of the cytosine derivative by the base excision **DNA repair** pathway (Gray 2015).

During the first intensive wave of DNA methylation very early in mammalian embryonal development, this process can be influenced by environmental stress impinging on the mother-to-be. It can be hunger stress, which is considered one of the most severe and fundamental stress experiences, but also psychological stress and stress caused by major life events (see Chap. 6). The process can also be influenced by dietary vitamins that are precursors of the methyl donating coenzyme, S-adenosyl methionine (Waterland and Jirtle 2003; Waterland and Jirtle 2004; Waterland et al. 2006). Research over the last 30 years has clearly shown that this epigenetic influence can have severe postnatal consequences, which are manifest throughout the whole of life. Moreover, there exists evidence that the intensive interaction of the child with the mother, or another primary caregiver, (or the lack thereof) also has a profound effect on disease susceptibility during the entire life of the person. In this case, as in others, epigenetic changes are very probable.

We will present here a discussion of some of the most studied epigenetic processes caused by pre- and postnatal stress of various origins.

These are:

- Undernutrition in utero (a much-investigated topic, to be found, for instance, in the literature on the Dutch Hunger Winter)
- Telomere shortening in mothers caring for severely sick children

- Maternal psychological stress during pregnancy
- Psychological stress and certain types of heart disease

In all of these cases, indirect and epidemiological studies from human medical history were the starting point, which was later reinforced by molecular biological and experimental work, mostly in rodents.

We will also include some of the most prevalent diseases for which a causative action, specifically related to the psychological and environmental stress of poverty, has been discussed in the literature.

In Utero-Acquired Predisposition for Metabolic Disease: The Example of the Dutch Hunger Winter

A very useful and up-to-date review dealing with the topic of the Dutch Hunger Winter in 1944/45 appeared in 2020 (Deodati et al. 2020). In the final phase of World War II with Allied forces already occupying large parts of France, the population of the Netherlands started an uprising and strike that was answered by the German army with severe measures against the population of the German-occupied western Netherlands, including a massive food embargo. This situation lasted from November 1944 until July 1945 (Burger et al. 1948), when the reestablished civilian government finally succeeded in supplying the area with an adequate amount of food. During the hunger winter registries and health care remained intact so that individuals exposed to massive undernutrition could be searched for by birth weight and other health parameters, and later for diseases during their adult lives and also into the next generation (Heijmans et al. 2008; Stein and Lumey 2000). Intensive studies of the cohort of survivors revealed correlations with many medical and socioeconomic parameters (Scholte et al. 2015; Lumey et al. 2007, 2011). Also, a clear correlation with negative labor market outcomes such as annual earnings and employment was found (Scholte et al. 2015). We are concentrating here on the work showing effects on DNA methylation that were stable for decades, when the members of the vulnerable group were about 60 years old (Heijmans et al. 2008; Tobi et al. 2018). One of the genomic regions studied was the **differentially methylated region (DMR)** of the **IGF2 (insulin-like growth factor 2)** gene promoter (Fig. 5.1). The IGF2 gene is maternally imprinted and codes for the Igf2 protein, a major regulator of growth and metabolism in mammals. In cases where severe malnutrition was experienced

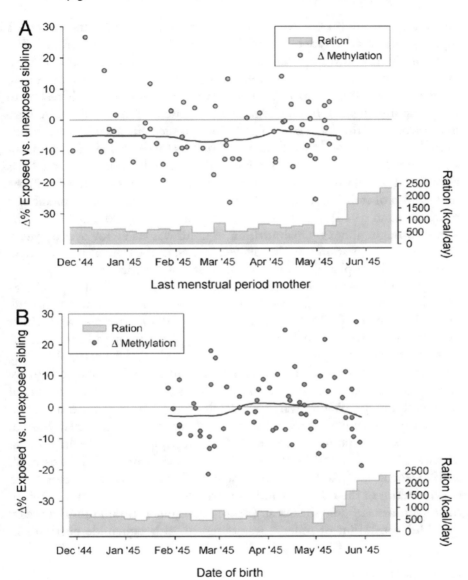

Fig. 5.1 Effects of extreme hunger in utero on DNA CpG methylation. The graphs show the differences of DNA CpG methylation between babies born in 1945 exposed in utero to the extreme hunger of the Dutch Hunger Winter and their unexposed siblings. Shown is CpG methylation of the differentially methylated promoter region of the human IGF2 gene. The methylation analysis was performed on leukocyte DNA about 6 decades after the hunger winter. Part **a** analyzes people who experienced the extreme hunger situation in utero periconceptually, here equated with the last menstrual period of the mother. As the graph shows, in these individuals the IGF2 promoter

periconceptionally (in the earliest days of gestation), the authors observed a highly significant decrease of about 5% in methylation of the DMR of IGF2. Controls in this investigation were same-sex siblings who had not been exposed to malnutrition in utero since they had been born 1 or 2 years earlier. They were very useful controls because they were genetically closely related to the group investigated. In individuals who experienced malnutrition in utero in the second or third trimester of gestation, no such effect could be seen. We make four comments here to explain and discuss these findings: (1) the sensitive period of the embryo was obviously the preimplantation period (as has been noted above, this is the period when epigenetic marks are erased); (2) the same group of individuals showed a highly significant increase in **metabolic syndrome** and type II diabetes later in life (Lumey et al. 2007, 2011); (3) the change brought about by malnutrition in utero was stable for over 60 years; (4) the gene in question was already well known at the time of the investigation and could be plausibly linked to the metabolic syndrome.

Very similar results were obtained and extended in animal experiments with rats exposed to malnutrition in utero (Burdge et al. 2007; Benyshek et al. 2006; Deodati et al. 2018; Thompson et al. 2010). The rats were either exposed to a general but balanced low-nutrition diet through ligation of the uterine artery or through the diet given to the mother. The diets applied were a general malnutrition or an isocaloric but protein-deficient diet (Burdge et al. 2007; Lumey et al. 2011). The results described above for survivors of the Dutch Hunger Winter were very clearly confirmed. In the present examples, the genes coding for transcription regulators and metabolic enzymes of the liver were hypomethylated and therefore upregulated, leading to an increase in **PEPCK** and in the **glucocorticoid receptor** 1_{10} (Morris et al. 2009; Deodati et al. 2020). The phenotypic effect is an increased response to stress in terms of the **HPA axis** and an increased response to fat in the diet. In the experiments of Benyshek et al. (2006), **transgenerational inheritance** by the third generation of rodents was observed, although undernutrition in

Fig. 5.1 (continued)

region is significantly under-methylated, leading to higher expression of Igf2. Part **b** shows babies born between February and June 1945. In utero, they experienced extreme hunger in the later part of pregnancy rather than in the early period. In these cases, there was no significant effect on CpG methylation on the promoter of IGF2. The effect seen in part a leads to a lifelong increase in IGF2 expression and to a significantly higher incidence of metabolic syndrome and type II diabetes in those individuals. The green curve at the bottom of the figure shows the daily calorie intake in the Western Netherlands in 1945. Source: Heijmans et al. (2008) with permission. © (2008) National Academy of Sciences, U.S.A.

pregnancy was only applied in the first generation. Another study showed a direct consequence for gene expression in **pancreatic islet cells**, and therefore for the generation of diabetes (Thompson et al. 2010).

A plausible interpretation for the fact that protein restriction brings about a similar effect to malnutrition is that protein restriction means a relative lack of methionine, which is a major methyl donor, and this indirectly leads to a lack of DNA methylation. Feeding mice with methionine and other methyl donors as well as with folic acid (a coenzyme indirectly involved in the DNA methylation reaction) has been shown to reverse the effects of undermethylation and the associated outcomes in term of disease predisposition (Waterland and Jirtle 2004). Of note, not only undernutrition in utero but also maternal diabetes and its oversupply of glucose led to an increased risk of metabolic syndrome and type II diabetes in the offspring (Boney et al. 2005; Dabelea et al. 2000).

Another highly interesting finding was that treating rats during pregnancy with **glucocorticoids** like **dexamethasone** (all other things being equal) had definite and strong effects on the health of the offspring and presumed epigenetic changes similar to the ones just described (Drake et al. 2005) as well as on disease predisposition phenotypes like hyperglycemia and hyperinsulinemia and also cardiometabolic disease in the second-generation offspring (Drake et al. 2005). We speak of presumed epigenetic changes here because DNA methylation or other chromatin changes in this experiment were not studied.

In the same vein, mice were treated with dexamethasone and a strong and significant down-regulation of **CD8 T cells** was induced, reducing the animals' defenses against tumors and infections (Hong et al. 2020). In both glucocorticoid experiments, the dose was given to the pregnant dams in the last part of gestation and perinatally. Thus, the effect is different from the one observed on IGF2 promoter methylation during the Dutch Hunger Winter, but nevertheless a strong and important one.

Environmental toxins like vinclozolin (an anti-androgen) and methoxychlor (an estrogenic compound) were revealed to have transgenerational epigenetic negative effects on the fertility of the male progeny even in the F4 generation (Anway et al. 2005). Vinclozolin is an antifungal agent used in agriculture (mainly in the wine industry) and methoxychlor is a pesticide that was used in the United States instead of the banned DDT, but in 2003 was also banned. Both substances are found in the environment, and both substances have strong toxic effects if ingested by humans, the one as an anti-androgen, the other as a synthetic estrogen. The investigations show that vinclozolin, by causing increasing CpG methylation at certain specific CpG islands, can cause reduced or absent male fertility in the **F1**, and interestingly, if those males are outcrossed to untreated females, similar strong male fertility defects are found in F2–F4. The effects can only be explained if in the male primordial germ cells maintenance methylase restores the epigenetic mark to both strands of the genomic DNA.

The transgenerational inheritance of epigenetic marks was observed in the female as well as in the male **germ-line** (Gluckman et al. 2007). Similar transgenerational inheritance was also observed in the grandchildren of those mothers who had survived the Dutch Hunger Winter. In these cases, birth weights were statistically analyzed, showing a clear correlation between maternal and offspring birth weight depending on the time of exposure to famine, but changes in DNA methylation were not studied (Stein and Lumey 2000).

When trying to explain these observations in terms of the epigenetic mechanisms involved, it must be assumed that the erasure of DNA methylation in the primordial germ-line, which is well established (Reik et al. 2001), was not completely executed in the gene promoters responsible for the expression of the genes mentioned above, which are responsible for the disease phenotypes studied. Another possible explanation is, of course, that other epigenetic marks besides CpG methylation could exist, escaping erasure and being transmitted transgenerationally.

Summarizing the results of the research on the Dutch Hunger Winter and on similar famines (the Siege of Leningrad, the Great Leap Forward famine in China, and others not discussed in detail here) and of the rodent experiments, it seems to be clear that the influence of undernutrition, (un)balanced diet, environmental toxins, and stress hormones during pregnancy exerts a strong effect on the health of the next generation. We can only speculate why these epigenetic mechanisms arose during biological evolution. Evolutionary biologists posit that in the evolution of mankind, the primary response of the fetus to a stressful situation—above all, of course, to the stress of hunger—was an adaptive response. The changes described in the metabolic system, primarily in the liver, result in some improvement of survival of the progeny under conditions of continued famine due to the more intensive and thrifty use of glucose and other basic components of food (Gluckman et al. 2009; Godfrey et al. 2007). The negative consequences (metabolic syndrome, cardiovascular disease, diabetes, obesity) become visible only if the progeny grows up with adequate or increased food supply, and only later in life.

Similar strong effects of diet and stress of various kinds exist during the 1st years of children's lives. One factor is the care given by the mother or another primary caregiver. As shown by Francis et al. (1999), a rat's lack of maternal care in newborn pups can lead to transgenerational inheritance of stress reactivity and fearfulness. Without analyzing DNA CpG methylation, the authors nevertheless showed clearly that maternal care (licking, grooming, and nursing) or the lack thereof can lead to a strong transgenerational effect on the behavior of F1 animals.

The dietary and all the other stress situations described above as causing transgenerational epigenetic changes in the individual are highly likely to occur as a consequence of poverty.

Telomere Maintenance and Shortening

In 2004, a landmark paper appeared in the PNAS describing the influence on the telomere length of severe stress in mothers caring for chronically ill children (Epel et al. 2004). The magnitude of this effect depended on the duration of the caring period. A paper that appeared 11 years later (Chen et al. 2015) confirmed these results and extended it to children in Chile with disabilities. A meta-analysis of papers correlating stress and telomere length (Mathur et al. 2016) strongly corroborated the general connection between perceived psychological stress and telomere attrition. Short **telomeres** are an epigenetic mark, because they are not restored during the life of an individual.

Telomere biology and in particular the attrition (shortening) of telomeres has been discussed extensively as a possible cause for aging, for certain kinds of disease (heart disease, neuropsychiatric diseases, and others), and as a primary sensor for chronic stress (Hornsby 2006; Aubert and Landsdorp 2008; Shay and Wright 2019; Mayer et al. 2019; Louzon et al. 2019).

Telomeres are protective **caps** on both ends of the linear eukaryotic chromosomes that consist of a highly repetitive DNA sequence (5'TTAGGG3' in humans and all vertebrates), a G-rich single strand 3' overhang, and a relatively large number of telomere-binding proteins, which stabilize the structure and prevent the cellular repair systems recognizing the ends of the chromosomes as double-strand breaks. If telomeres become critically short (the exact repeat number which constitutes "critically short" is known only in specialized cases; Lin et al. 2014) or if certain telomeric proteins are missing (for instance through a **Mendelian mutation** in one of the telomere-binding proteins), the telomeres are "repaired" by nonhomologous end joining or recombinational repair that can lead to cell death and in some cases to highly **aneuploid** cells within a few cell generations. The process is frequently observed in cancer cells.

The repeated hexamer sequence mentioned above cannot be faithfully replicated by semiconservative replication at the extreme end of the chromosome due to the need to create a primer from which the replicative DNA polymerase can start synthesizing the new strand. This means that the 5' ends of the single DNA strands at both ends of the chromosome become shorter with every replication round. In the absence of **telomerase**, a reverse transcriptase that can synthesize telomeres de novo, chromosome ends become shorter with every replication round; this is the cause for crisis and cellular senescence in the cell cultures of primary human cells, a process also referred to as "Hayflick aging" (Hayflick and Moorhead 1961; Olovnikov 1973).

In 2009, the Nobel Prize in Physiology or Medicine was awarded to Elizabeth Blackburn (the co-author of the landmark paper from 2004 mentioned above) together with Carol Greider and Jack Szostak for the discovery and functional analysis of telomeres.

The relationship between telomere maintenance and organismic aging is far from clear even today (Hornsby 2006; Aubert and Landsdorp 2008; Frenck et al. 1998; Mensà et al. 2019; Blackburn et al. 2015). We mention here some of the open questions, for which no answers or only controversial answers are available. For practical reasons, most of the determinations of telomere length (e.g., in Epel et al. 2004) were done on **leukocytes**; does this correlate well with other tissues that may be more relevant for aging? The telomere length that is critical for cellular functioning in vivo in humans is unknown. As mentioned above, a minimum length is known only in the case of avoiding nonhomologous end fusion at 3.8 kb in lymphocytes in **chronic lymphocytic leukemia** (Lin et al. 2014). Telomere length seems to be very unevenly distributed in cells in a given tissue and between different chromosomes in one cell (Aubert and Landsdorp 2008). Is telomere shortening in a minority of the chromosomes of a cell critical for cellular function? Most importantly, is the phenomenon of Hayflick aging in culture relevant for the aging process of the whole organism?

Cellular senescence occurs in organs of aged individuals even in the absence of telomere shortening, as shown by many species whose somatic cells are postmitotic, like *C. elegans* and *D. melanogaster* (Blackburn et al. 2015). However, selective killing of senescent cells identified by senescence markers such as **senescence-associated beta-galactosidase (SA beta-galactosidase)** does have an effect on rejuvenation (Baar et al. 2017). Telomere shortening, even though it may not cause cellular senescence in vivo, may be a mark of pathological and disease processes caused by severe and chronic psychological stress (Epel et al. 2004). We are therefore describing here the results of the landmark paper by Epel, Blackburn, and colleagues and of the papers investigating this phenomenon further, including those focused on the mother's prenatal stress acting on the fetus and experiments performed with rodents.

The key finding of the landmark paper (Epel et al. 2004) is that mothers caring for severely ill children show progressively shorter telomeres in the leukocytes, depending on the duration of this perceived stress, even after correcting for age and other possible confounding factors.

Despite all the well-founded skepticism concerning telomeres as a cause of aging, there can be no doubt that a disturbance of telomeres can lead to very severe diseases like *dyskeratosis congenita* (Vulliamy et al. 2001; Holohan et al. 2014; Calado and Young 2009). In this cluster of systemic diseases, organs like skin, nails, and bone marrow are affected, all of which are active in cell division (and therefore experience telomere shortening) throughout life. In many cases, this leads to death, e.g., in **aplastic anemia**. Importantly, the shortening of telomeres that has been observed is also strongly dependent on environmental factors (Calado and Young 2009; von Zglinicki 2002; Mirabello et al. 2009) and environmental pollution (Senthilkumar et al. 2011, 2012; Dioni et al. 2011). This is the case even in **idiopathic disease** without any family history or a responsible genetic mutation known of (Zhang et al. 2013). Environmental factors include toxic environmental substances, noise (Maschke et al. 2000), and psychological stress. It appears that telomeres are very susceptible to chronic stress (Mayer et al. 2019), and the first paper by Epel and colleagues as well as the papers that followed corroborate this (Epel et al. 2004, 2006, 2009a, b).

The effect of severe stress on telomere length is an example of stress causing epigenetic changes. It is unknown if the epigenetic process of telomere attrition is related to changes in the methylation pattern. There is no evidence for changes in the CpG methylation pattern reported in the papers by Epel, Blackburn, and colleagues. But it is known that short telomeres can be caused by adverse events, such as stress during pregnancy and early childhood (Entringer et al. 2011; Price et al. 2013), and can even be epigenetically inherited by the next generation (Collopy et al. 2015).

Epigenetics, Psychological Stress, Heart Disease, and Further Observations on Stress-Induced Diseases

As we have already seen in the previous chapter, stress exposure can directly lead to critical physiological incidents such as stress-induced cardiomyopathy, also known as **Takotsubo disease** or broken heart syndrome, indicating severe emotional stress as the most relevant trigger, (Münzel et al. 2016; Schneider 2012; see also Chap. 4 and Fig. 4.2a–c). Heart disease also figures prominently when we investigate the field of stress-related epigenetic changes and resulting detrimental health outcomes.

Coronary heart disease (CHD) is statistically the most prominent cause of death in countries all over the world, including EU member nations and the United States. CHD includes among other clinical syndromes angina pectoris, heart infarction, and also, indirectly, stroke, which can be caused by thrombosis originating in the cardiovascular system. In the studies of the Dutch Hunger Winter survivors described above, epigenetic modifications of the genome were found to have a strong influence on CHD, in parallel with other diseases found in the survivors, such as metabolic syndrome and type II diabetes (Heijmans et al. 2008; Painter et al. 2006). These epigenetic modifications, as we have seen, are acquired during a stressful life in utero.

After birth and during adult life, chronic stress of various kinds, including the experience of poverty, is another strong inducer of CHD (Lagraauw et al. 2015; Gluckman et al. 2009). The clinical descriptions of CHD are often summarized in the literature under the inclusive term "ischemic heart disease," which can range from a mild defect in blood flow (**ischemia**) in the heart muscle to a life-threatening disease. The physiological causes for CHD are well researched and can be summarized under the heading of "lifestyle." The lifestyle factors that cause CHD are, for example, convincingly summarized and correlated with a decrease in life expectancy in the ongoing large-scale "Nurses' Health Study," begun in 1980, and the subsequent "Health Professionals Follow-up Study" (Li et al. 2018). The most important lifestyle factors causing CHD are smoking, alcohol consumption, a **body mass index** over 25 kg/m^2, physical activity under 30 min/day, and a poor-quality diet. The combination of these factors, depending on the amount (weight) of each of the factors, can lead on average to a difference in overall life expectancy of about 15 years (Li et al. 2018). Among the proximate causes found as a result of an unhealthy lifestyle are increases in serum cholesterol and LDL serum lipids, **atherosclerosis**, diabetes, metabolic syndrome. Joining these accepted contributors to heart disease, an increasing number of **stressors** are also recognized today, which together with lifestyle and genetic predisposition factors contribute to the timing and severity of CHD.

Some of the psychological and social stress factors contributing to CHD are closely correlated with the stress of poverty (Sundquist et al. 2004), and some also closely interact with the lifestyle factors just mentioned. The psychological stress factors include (in this order) death of partner or spouse, death of a closely related person, break up of a relationship or divorce, learning of a cancer diagnosis or other severe illness, business failure, losing a job, living without a job for a prolonged time, bullying in the workplace, and unstable family situations, among others (Lagraauw et al. 2015; Rosengren et al. 2004). It is important to mention that both singular adverse life events (such as the death of a partner) and prolonged chronic stress (such as chronic noise exposure or long-term unemployment, see below) can have a large effect on the timing and severity of CHD. The psychological and social stress factors are now well known and are published and distributed to the general public by institutions such as the German Heart Foundation (Ladwig et al. 2016; see also Li et al. 2019). The close correlation of CHD with poverty is convincingly shown in a number of longitudinal health studies such as the ones just mentioned; additional longitudinal studies in the United States and 11 EU countries are reviewed and referenced in Sundquist et al. (2004).

Psychosocial stress factors make a significant contribution to acute ischemic heart disease, both in patients with already increased physiological risk factors such as atherosclerosis and also in patients who are essentially free of the physiological risk factors such as a blocked coronary artery.

The correlation of poverty or **socioeconomic status** (low **SES**) with mortality, disease-specific mortality (in the present case, CHD), and the incidence of ischemic heart disease requires long-term studies, careful definition of income and education variables, and careful investigation of what constitutes a "poor neighborhood" (Bucher and Ragland 1995; Kunst et al. 1999; Hemingway et al. 2000; Salomaa et al. 2000; Wamala 2001). Among other environmental indicators, noise in particular is a very important component of a poor neighborhood (Münzel et al. 2014, 2016). The effects of noise stress are described in more detail in a series of papers from Germany (Maschke et al. 2000; Münzel et al. 2016; Babisch 2000).

Another recent cross-sectional study was performed in the United States between 2011 and 2014 and published in a series of papers (Sullivan et al. 2018; Hammadah et al. 2017a, b, 2018). The **Mental Stress Ischemia Prognosis Study** (**MIPS**) encompassed 695 patients and measured stress reactivity (in terms of **adrenal hormone** response) in combination with blood pressure, heart rate, rate-pressure product, and inflammatory markers such as **IL-6**, monocyte chemoattractant protein 1 (MCP-1), matrix metalloproteinase 9 (MMP-9), and C-reactive protein (CRP). The participants were selected for previous mild ischemic disease but excluded for recent acute events such as myocardial infarction. The information about "poor neighborhood" (or otherwise, "good neighborhood") was taken from residential addresses geocoded and merged with poverty data from the American Community Survey (Sullivan et al. 2019).

The MIPS, although not a **longitudinal study**, produced a number of unexpected and surprising data, which in retrospect show a consistent pattern. The patients underwent tests for mental stress (speaking in front of an audience), pharmacologically induced stress (not discussed in detail here), and physical stress (ergometer or treadmill type). The reaction of the patients to mental stress led to an increase in heart rate, three inflammatory markers (but not CRP), myocardial ischemia, and stress hormone concentration, in particular **epinephrine**. Serum epinephrine was measured at rest and two minutes after the mental stress (Sullivan et al. 2018). For reasons that will be described in Chap. 6, epinephrine seems to be better suited here than **cortisol**, because it does not show the large diurnal variations found for cortisol (Dijckmans et al. 2017). The reaction to mental stress with respect to hemodynamic parameters and epinephrine concentrations was blunted significantly in those patients coming from a poor neighborhood, and the recovery phase took longer in these patients (Sullivan et al. 2019). These changes were not seen if the stress resulted from physical activity, as opposed to from a mental stress test. It is unknown at present how a blunted reaction to mental stress can be correlated causally with the higher incidence of CHD in the probands coming from poor neighborhoods. However, what we can say at present is that a dynamic, fully developed hormonal stress response seems to be important for a healthy life.

In a further investigation following MIPS (Hammadah et al. 2018), mental stress tests were combined with the direct measurement of the influence of the stress on myocardial ischemia by using **Tc-99m** injection and a radioactive imaging camera to detect blood flow in the heart muscle. The mental stress always induced the now well-known changes in hemodynamic parameters and in inflammation markers but did not seem to be clearly correlated with producing myocardial ischemia. It is probable that to induce myocardial ischemia additional unknown prerequisites are necessary; these prerequisites may turn out to be additional risk factors for CHD. Many more tests and physiological measurements will be necessary to give a complete interpretation of these findings. However, even now we can say that a normal, healthy stress reaction requires both a higher peak and a more rapid recovery phase. What is being compared here are people living in high-poverty and low-poverty neighborhoods, with both groups suffering from chronic, non-acute cardiac ischemia (Sullivan et al. 2019). The results seem to support the notion that chronic stress, and thus also the chronic stress of poverty, can lead to a greater risk of developing coronary heart disease and could therefore explain the disease and mortality pattern found in the longitudinal studies (Sundquist et al. 2004).

It is particularly interesting that three of the inflammatory markers (IL-6, MMP-9, and MCP-1) measured after mental stress always increased but less so in patients coming from poor neighborhoods. These results parallel those concerning epinephrine. There is presently no explanation, but we find it probable that the results have to do with the possibility that both IL-6 and other inflammatory markers have other still unknown functions in relation to stress besides indicating an acute inflammation. IL6 and other **cytokines** of the human immune system act in a complex network and depending on context can present both pro-inflammatory and anti-inflammatory activity (Scheller et al. 2011). It is also worth considering the role of IL-6 in the currently popular hypothesis of "inflammaging" (Franceschi et al. 2017), in which chronic low-grade inflammation seems to be important for the aging process in the absence of acute inflammation.

In conclusion, what we have tried to explain and describe in detail in this chapter is that epigenetic changes in the genome of fetuses and young children can lead to characteristic diseases or disease preconditioning many years later, even in subsequent generations. We have done so by drawing on studies investigating two prominent examples, namely: (1) the consequences of the Dutch Hunger Winter, leading to metabolic syndrome and related diseases, and (2) telomere shortening in leukocytes of single mothers caring for severely sick children. In addition, we compiled evidence from clinical and longitudinal studies for the strong correlation of environmental stress, including the stress of poverty, with disease incidence.

References

Anway MD, Cupp AS, Uzumcu M, Skinner MK (2005) Epigenetic transgenerational actions of endocrine disruptors and male fertility. Science 308(5727):1466–1469. https://doi.org/10.1126/science.1108190

Aubert G, Landsdorp PM (2008) Telomeres and aging. Physiol Rev 88(2):557–579. https://doi.org/10.1152/physrev.00026.2007

Baar MP, Brandt RMC, Putavet DA, Klein JDD, Derks KWJ, Bourgeois BRM, Stryeck S, Rijksen Y, von Willigenburg H, Feijtel DA, van der Pluijm I, Essers J, van Cappellen WA, von IJcken WF, Houtsmuller AB, Pothof J, de Bruin RWF, Madl T, Hoeijmakers JHJ, Campisi J, de Keizer PLJ (2017) Targeted apoptosis of senescent cells restores tissue homeostasis in response to chemotoxicity and aging. Cell 169(1):132–147. https://doi.org/10.1016/j.cell.2017.02.031

Babisch W (2000) Traffic noise and cardiovascular disease. Epidemiological review and synthesis. Noise Health 2(8):9–32

Benyshek DC, Johnston CS, Martin JF (2006) Glucose metabolism is altered in the adequately-nourished grand-offspring (F$_3$ generation) of rats malnourished during gestation and perinatal life. Diabetologia 49:1117–1119. https://doi.org/10.1007/s00125-006-0196-5

Blackburn EH, Epel ES, Lin J (2015) Human telomere biology: a contributory and interactive factor in aging, disease risks, and protection. Science 350(6265):1193–1198. https://doi.org/10.1126/science.aab3389

Boney CM, Verma A, Tucker R, Vohr BR (2005) Metabolic syndrome in childhood: association with birth weight, maternal obesity, and gestational diabetes mellitus. Pediatrics 115(3):e290–e296. https://doi.org/10.1542/peds.2004-1808

Bucher HC, Ragland DR (1995) Socioeconomic indicators and mortality from coronary heart disease and cancer: a 22-year follow-up of middle-aged men. Am J Public Health 85(9):1231–1236. https://doi.org/10.2105/ajph.85.9.1231

Burdge GC, Slater-Jefferies J, Torrens C, Phillips ES, Hanson MA, Lillycrop KA (2007) Dietary protein restriction of pregnant rats in the F_0 generation induces altered methylation of hepatic gene promoters in the adult male offspring in the F_1 and F_2 generations. Br J Nutr 97(3):435–439. https://doi.org/10.1017/S0007114507352392

Burger GCE, Drummond JC, Sandstead HR (1948) Malnutrition and starvation in Western Netherlands, September 1944 – July 1945, Part I. General State Printing Office, The Hague

Calado RT, Young NS (2009) Telomere diseases. N Engl J Med 361(24):2353–2365. https://doi.org/10.1056/NEJMra0903373

Chen X, Velez JC, Barbosa C, Pepper M, Andrade A, Stoner L, De Vivo I, Gelaye B, Williams MA (2015) Smoking and perceived stress in relation to short salivary telomere length among caregivers of children with disabilities. Stress 18(1):20–28. https://doi.org/10.3109/10253890.2014.969704

Collopy LC, Walne AJ, Cardoso S, de la Fuente J, Mohamed M, Toriello H, Tamary H, Ling AJYV, Lloyd T, Kassam R, Tummala H, Vulliamy T, Dokal I (2015) Triallelic and epigenetic-like inheritance in human disorders of telomerase. Blood 126(2):176–184. https://doi.org/10.1182/blood-2015-03-633388

Dabelea D, Knowler WC, Pettitt DJ (2000) Effect of diabetes in pregnancy on offspring: follow-up research in the Pima Indians. J Matern Fetal Med 9:83–88

Deodati A, Argemi J, Germani D, Puglianiello A, Alisi A, De Stefanis C, Ferrero R, Nobili V, Aragón T, Cianfarani S (2018) The exposure to uteroplacental insufficiency is associated with activation of unfolded protein response in postnatal life. PLoS One 13(6):1–14. https://doi.org/10.1371/journal.pone.0198490

Deodati A, Inzaghi E, Cianfarani S (2020) Epigenetics and in utero acquired predisposition to metabolic disease. Front Genet 10:1270. https://doi.org/10.3389/fgene.2019.01270

Dijckmans B, Tortosa-Martínez J, Caus N, González-Caballero G, Martínez-Pelegrin B, Manchado-Lopez C, Cortell-Tormo JM, Chulvi-Medrano I, Clow A (2017) Does the diurnal cycle of cortisol explain the relationship between physical performance and cognitive function in older adults? Eur Rev Aging Phys Act 14(6):1–10. https://doi.org/10.1186/s11556-017-0175-5

Dioni L, Hoxha M, Nordio F, Bonzini M, Tarantini L, Albetti B, Savarese A, Schwartz J, Bertazzi PA, Apostoli P, Hou L, Baccarelli A (2011) Effects of short-term exposure to inhalable particulate matter on telomere length, telomerase expression, and telomerase methylation in steel workers. Environ Health Perspect 119(5):622–627. https://doi.org/10.1289/ehp.1002486

Drake AJ, Walker BR, Seckl JR (2005) Intergenerational consequences of fetal programming by in utero exposure to glucocorticoids in rats. Am J Physiol Regul Integr Comp Physiol 288(1):R34–R38. https://doi.org/10.1152/ajpregu.00106.2004

Entringer S, Epel ES, Kumsta R, Lin J, Hellhammer DH, Blackburn EH, Wüst S, Wadhwa PD (2011) Stress exposure in intrauterine life is associated with shorter telomere length in young adulthood. Proc Natl Acad Sci 108(33):E513–E518. https://doi.org/10.1073/pnas.1107759108

Epel ES, Blackburn EH, Lin J, Dhabhar FS, Adler NE, Morrow JD, Cawthon RM (2004) Accelerated telomere shortening in response to life stress. Proc Natl Acad Sci USA 101(49):17312–17315. https://doi.org/10.1073/pnas.0407162101

Epel ES, Lin J, Wilhelm FH, Wolkowitz OM, Cawthon R, Adler NE, Dolbier C, Mendes WB, Blackburn EH (2006) Cell aging in relation to stress arousal and cardiovascular disease risk factors. Psychoneuroendocrinology 31(3):277–287. https://doi.org/10.1016/j.psyneuen.2005.08.011

Epel ES, Stein Merkin S, Cawthon R, Blackburn EH, Adler NE, Pletcher MJ, Seeman TE (2009a) The rate of leukocyte telomere shortening predicts mortality from cardiovascular disease in elderly men. Aging 1(1):81–88. https://doi.org/10.18632/aging.100007

Epel E, Daubenmier J, Moskowitz JT, Folkman S, Blackburn E (2009b) Can meditation slow rate of cellular aging? Cognitive stress, mindfulness, and telomeres. Ann N Y Acad Sci 1172(1):34–53. https://doi.org/10.1111/j.1749-6632.2009.04414.x

Franceschi C, Garagnani P, Vitale G, Capri M, Salvioli S (2017) Inflammaging and 'Garb-aging'. Trends Endocrinol Metab 28(3):199–212. https://doi.org/10.1016/j.tem.2016.09.005

Francis D, Diorio J, Liu D, Meaney MJ (1999) Nongenomic transmission across generations of maternal behavior and stress responses in the rat. Science 286(5442):1155–1158. https://doi.org/10.1126/science.286.5442.1155

Frenck RW, Blackburn EH, Shannon KM (1998) The rate of telomere sequence loss in human leukocytes varies with age. Proc Natl Acad Sci U S A 95:5607–5610. https://doi.org/10.1073/PNAS.95.10.5607

Gluckman PD, Hanson MA, Beedle AS (2007) Non-genomic transgenerational inheritance of disease risk. Bioessays 29(2):145–154. https://doi.org/10.1002/bies.20522

Gluckman PD, Hanson MA, Buklijas T, Low FM, Beedle AS (2009) Epigenetic mechanisms that underpin metabolic and cardiovascular diseases. Nat Rev Endocrinol 5:401–408. https://doi.org/10.1038/nrendo.2009.102

Godfrey KM, Lillycrop KA, Burdge GC, Gluckman PD, Hanson MA (2007) Epigenetic mechanisms and the mismatch concept of the developmental origins of health and disease. Pediatr Res 61(5):5R–10R. https://doi.org/10.1203/pdr.0b013e318045bedb

Goldberg AD, Allis CD, Bernstein E (2007) Epigenetics: a landscape takes shape. Cell 128(4):635–638. https://doi.org/10.1016/j.cell.2007.02.006

Goll MG, Bestor TH (2005) Eukaryotic cytosine methyltransferases. Annu Rev Biochem 74:481–514. https://doi.org/10.1146/annurev.biochem.74.010904.153721

Gray SG (ed) (2015) Epigenetic cancer therapy. Academic, Oxford

Hammadah M, Al Mheid I, Wilmot K, Ramadan R, Shah AJ, Sun Y, Pearce B, Garcia EV, Kutner M, Bremner JD, Esteves F, Raggi P, Sheps DS, Vaccarino V, Quyyumi AA (2017a) The Mental Stress Ischemia Prognosis Study (MIPS): objectives, study design, and prevalence of inducible ischemia. Psychosom Med 79(3):311–317. https://doi.org/10.1097/PSY.0000000000000442

Hammadah M, Alkhoder A, Al Mheid I, Wilmot K, Isakadze N, Abdulhadi N, Chou D, Obideen M, O'Neal WT, Sullivan S, Tahhan AS, Kelli HM, Ramadan R, Pimple P, Sandesara P, Shah AJ, Ward L, Ko Y, Sun Y, Uphoff I, Pearce B, Garcia EV, Kutner M, Bremner JD, Esteves F, Sheps DS, Raggi P, Vaccarino V, Quyyumi AA (2017b) Hemodynamic, catecholamine, vasomotor and vascular responses: determinants of myocardial ischemia during mental stress. Int J Cardiol 243:47–53. https://doi.org/10.1016/j.ijcard.2017.05.093

Hammadah M, Sullivan S, Pearce B, Al Mheid I, Wilmot K, Ramadan R, Tahhan AS, O'Neal WT, Obideen M, Alkhoder A, Abdelhadi N, Kelli HM, Ghafeer MM, Pimple P, Sandesara P, Shah AJ, Hosny KM, Ward L, Ko Y, Sun YV, Weng L, Kutner M, Bremner JD, Sheps DS, Esteves F, Raggi P, Vaccarino V, Quyyumi AA (2018) Inflammatory response to mental stress and mental stress induced myocardial ischemia. Brain Behav Immun 68:90–97. https://doi.org/10.1016/j.bbi.2017.10.004

Hayflick L, Moorhead PS (1961) The serial cultivation of human diploid cell strains. Exp Cell Res 25(3):585–621. https://doi.org/10.1016/0014-4827(61)90192-6

Heijmans BT, Tobi EW, Stein AD, Putter H, Blauw GJ, Susser ES, Slagboom PE, Lumey LH (2008) Persistent epigenetic differences associated with prenatal exposure to famine in humans. Proc Natl Acad Sci USA 105(44):17046–17049. https://doi.org/10.1073/pnas.0806560105

Hemingway H, Shipley M, Macfarlane P, Marmot M (2000) Impact of socioeconomic status on coronary mortality in people with symptoms, electrocardiographic abnormalities, both or neither: the original Whitehall study 25 year follow up. J Epidemiol Community Health 54(7):510–516. https://doi.org/10.1136/jech.54.7.510

Holohan B, Wright WE, Shaw JW (2014) Telomeropathies: an emerging spectrum disorder. J Cell Biol 205(3):289–299. https://doi.org/10.1083/jcb.201401012

Hong JY, Lim J, Carvalho F, Cho JY, Vaidyanathan B, Yu S, Annicelli C, Ip WKE, Medzhitov R (2020) Long-term programming of CD8 T cell immunity by perinatal exposure to glucocorticoids. Cell 180(5):847–861. https://doi.org/10.1016/j.cell.2020.02.018

Hornsby PJ (2006) Short telomeres: cause or consequence of aging? Aging Cell 5(6):577–578. https://doi.org/10.1111/j.1474-9726.2006.00249.x

Kunst AE, Groenhof F, Andersen O, Borgan J-K, Costa G, Desplanques G, Filakti H, Giraldes MdR, Faggiano F, Harding S, Junker C, Martikainen P, Minder C, Nolan B, Pagnanelli F, Regidor E, Vågerö D, Valkonen T, Mackenbach JP (1999) Occupational class and ischemic heart disease mortality in the United States and 11 European countries. Am J Public Health 89(1):47–53. https://doi.org/10.2105/ajph.89.1.47

Ladwig K-H, Münzel T, Meinertz T (2016) Psychischer und Sozialer Stress. Deutsche Herzstiftung eV., Frankfurt a. M.

Lagraauw HM, Kuiper J, Bot I (2015) Acute and chronic psychological stress as risk factors for cardiovascular disease: insights gained from epidemiological, clinical and experimental studies. Brain Behav Immun 50:18–30. https://doi.org/10.1016/j.bbi.2015.08.007

Li Y, Pan A, Wang DD, Liu X, Dhana K, Franco OH, Kaptoge S, Di Angelantonio E, Stampfer M, Willett WC, Hu FB (2018) Impact of healthy lifestyle factors on life expectancies in the US population. Circulation 138(4):435–355. https://doi.org/10.1161/circulationaha.117.032047

Li J, Atasoy S, Fang X, Angerer P, Ladwig K-H (2019) Combined effect of work stress and impaired sleep on coronary and cardiovascular mortality in hypertensive workers: the MONICA/KORA cohort study. Eur J Prev Cardiol 1–8. https://doi.org/10.1177/2047487319839183

Lin TT, Norris K, Heppel NH, Pratt G, Allan JM, Allsup DJ, Bailey J, Cawkwell L, Hills R, Grimstead JW, Jones RE, Britt-Compton B, Fegan C, Baird DM, Pepper C (2014) Telomere dysfunction accurately predicts clinical outcome in chronic lymphocytic leukaemia, even in patients with early stage disease. Br J Haematol 167(2):214–223. https://doi.org/10.1111/bjh.13023

Louzon M, Coeurdassier M, Gimbert F, Pauget B, de Vaufleury A (2019) Telomere dynamic in humans and animals: review and perspectives in environmental toxicology. Environ Int 131:105025. https://doi.org/10.1016/j.envint.2019.105025

Lumey LH, Stein AD, Kahn HS, van der Pal-de Bruin KM, Blauw GJ, Zybert PA, Susser ES (2007) Cohort profile: the Dutch Hunger Winter families study. Int J Epidemiol 36(6):1196–1204. https://doi.org/10.1093/ije/dym126

Lumey LH, Stein AD, Susser E (2011) Prenatal famine and adult health. Annu Rev Public Health 32:1–26. https://doi.org/10.1146/annurev-publhealth-031210-101230

Maschke C, Rupp T, Hecht K (2000) The influence of stressors on biochemical reactions – a review of present scientific findings with noise. Int J Hyg Environ Health 203(1):45–53. https://doi.org/10.1078/S1438-4639(04)70007-3

Mathur MB, Epel E, Kind S, Desai M, Parks CG, Sandler DP, Khazeni N (2016) Perceived stress and telomere length: a systematic review, meta-analysis, and methodologic considerations for advancing the field. Brain Behav Immun 54:158–169. https://doi.org/10.1016/j.bbi.2016.02.002

Mayer SE, Prather AA, Puterman E, Lin J, Arenander J, Coccia M, Shields GS, Slavich GM, Epel ES (2019) Cumulative lifetime stress exposure and leukocyte telomere length attrition: the unique role of stressor duration and exposure timing. Psychoneuroendocrinology 104:210–218. https://doi.org/10.1016/j.psyneuen.2019.03.002

Mensà E, Latini S, Ramini D, Storci G, Bonafè M, Olivieri F (2019) The telomere world and aging: analytical challenges and future perspectives. Ageing Res Rev 50:27–42. https://doi.org/10.1016/j.arr.2019.01.004

Messerschmidt DM, Knowles BB, Solter D (2014) DNA methylation dynamics during epigenetic reprogramming in the germline and preimplantation embryos. Genes Dev 28:812–828. https://doi.org/10.1101/gad.234294.113

Mirabello L, Huang W-Y, Wong JYY, Chatterjee N, Reding D, Crawford ED, De Vivo I, Hayes RB, Savage SA (2009) The association between leukocyte telomere length and cigarette smoking, dietary and physical variables, and risk of prostate cancer. Aging Cell 8(4):405–413. https://doi.org/10.1111/j.1474-9726.2009.00485.x

Morris TJ, Vickers M, Gluckman P, Gilmour S, Affara N (2009) Transcriptional profiling of rats subjected to gestational undernourishment: implications for the developmental variations in metabolic traits. PLoS One 4(9):e7271. https://doi.org/10.1371/journal.pone.0007271

Münzel T, Gori T, Babisch W, Basner M (2014) Cardiovascular effects of environmental noise exposure. Eur Heart J 35(13):829–836. https://doi.org/10.1093/eurheartj/ehu030

Münzel T, Knorr M, Schmidt F, von Bardeleben S, Gori T, Schulz E (2016) Airborne disease: a case of a Takotsubo cardiomyopathy as a consequence of nighttime aircraft noise exposure. Eur Heart J 37(37):2844. https://doi.org/10.1093/eurheartj/ehw314

Olovnikov AM (1973) A theory of marginotomy: the incomplete copying of template margin in enzymic synthesis of polynucleotides and biological significance of the phenomenon. J Theor Biol 41(1):181–190. https://doi.org/10.1016/0022-5193(73)90198-7

Painter RC, de Rooij SR, Bossuyt PM, Simmers TA, Osmond C, Barker DJ, Bleker OP, Roseboom TJ (2006) Early onset of coronary artery disease after prenatal exposure to the Dutch famine. Am J Clin Nutr 84(2):322–327

Peters J (2014) The role of genomic imprinting in biology and disease: an expanding view. Nat Rev Genet 15:517–530. https://doi.org/10.1038/nrg3766

Price LH, Kao H-T, Burgers DE, Carpenter LL, Tyrka AR (2013) Telomeres and early-life stress: an overview. Biol Psychiatry 73(1):15–23. https://doi.org/10.1016/j.biopsych.2012.06.025

Reik W, Dean W, Walter J (2001) Epigenetic reprogramming in mammalian development. Science 293(5532):1089–1093. https://doi.org/10.1126/science.1063443

Rosengren A, Hawken S, Ôunpuu S, Sliwa K, Zubaid M, Almahmeed WA, Ngu Blackett K, Sitthi-amorn C, Sato H, Yusuf S, For the INTERHEART Investigators (2004) Association of psychosocial risk factors with risk of acute myocardial infarction in 11 119 cases and 13 648 controls from 52 countries (the INTERHEART study): case-control study. Lancet 364:953–962

Salomaa V, Niemelä M, Miettinen H, Ketonen M, Immonen-Räihä P, Koskinen S, Mähönen M, Lehto S, Vuorenmaa T, Palomäki P, Mustaniemi H, Kaarsalo E, Arstila M, Torppa J, Kuulasmaa K, Puska P, Pyörälä K, Tuomilehto J (2000) Relationship of socioeconomic status to the incidence and prehospital, 28-day, and 1-year mortality rates of acute coronary events in the FINMONICA myocardial infarction register study. Circulation 101(16):1913–1918. https://doi.org/10.1161/01.CIR.101.16.1913

Scheller J, Chalaris A, Schmidt-Arras D, Rose-John S (2011) The pro- and anti-inflammatory properties of the cytokine interleukin-6. Biochim Biophys Acta 1813(5):878–888. https://doi.org/10.1016/j.bbamcr.2011.01.034

Schneider B (2012) Wenn das Herz zerbricht. Die Tako-Tsubo-Kardiomyopathie. HERZ HEUTE. Zeitschrift der Deutschen Herzstiftung e.V. 32(2):4–8

Scholte RS, van den Berg GJ, Lindeboom M (2015) Long-run effects of gestation during the Dutch Hunger Winter famine on labor market and hospitalization outcomes. J Health Econ 39:17–30. https://doi.org/10.1016/j.jhealeco.2014.10.002

Senthilkumar PK, Klingelhutz AJ, Jacobus JA, Lehmler H, Robertson LW, Ludewig G (2011) Airborne polychlorinated biphenyls (PCBs) reduce telomerase activity and shorten telomere length in immortal human skin keratinocytes (HaCat). Toxicol Lett 204(1):64–70. https://doi.org/10.1016/j.toxlet.2011.04.012

Senthilkumar PK, Robertson LW, Ludewig G (2012) PCB153 reduces telomerase activity and telomere length in immortalized human skin kerantinocytes (HaCaT) but not in human foreskin keratinocytes (NFK). Toxicol Appl Pharmacol 259(1):115–123. https://doi.org/10.1016/j.taap.2011.12.015

Shay JW, Wright WE (2019) Telomeres and telomerase: three decades of progress. Nat Rev Genet 20:299–309. https://doi.org/10.1038/s41576-019-0099-1

Stein AD, Lumey LH (2000) The relationship between maternal and offspring birth weights after maternal prenatal. Hum Biol 72(4):641–654

Sullivan S, Hammadah M, Al Mheid I, Wilmot K, Ramadan R, Alkhoder A, Isakadze N, Shah A, Levantsevych O, Pimple P, Kutner M, Ward L, Garcia EV, Nye J, Mehta PK, Lewis TT, Bremner JD, Raggi P, Quyyumi AA, Vaccarino V (2018) Sex differences in hemodynamic and microvascular mechanisms of myocardial ischemia induced by mental stress. Arterioscler Thromb Vasc Biol 38(2):473–480. https://doi.org/10.1161/ATVBAHA.117.309535

Sullivan S, Kelli HM, Hammadah M, Topel M, Wilmot K, Ramadan R, Pearce BD, Shah A, Lima BB, Kim JH, Hardy S, Levantsevych O, Obideen M, Kaseer B, Ward L, Kutner M, Hankus A, Ko Y-A, Kramer MR, Lewis TT, Brammer JD, Quyyumi A, Vaccarino V (2019) Neighborhood poverty and hemodynamic, neuroendocrine, and immune response to acute stress among patients with coronary artery disease. Psychoneuroendocrinology 100:145–155. https://doi.org/10.1016/j.psyneuen.2018.09.040

Sundquist K, Winkleby M, Ahlén H, Johansson S-E (2004) Neighborhood socioeconomic environment and incidence of coronary heart disease: a follow-up study of 25,319 women and men in Sweden. Am J Epidemiol 159(7):655–662. https://doi.org/10.1093/aje/kwh096

Thompson RF, Fazzari MJ, Niu H, Barzilai N, Simmons RA, Greally JM (2010) Experimental intrauterine growth restriction induces alterations in DNA methylation and gene expression in pancreatic islets of rats. J Biol Chem 258(20):15111–15118. https://doi.org/10.1074/jbc.M109.095133

Tobi EW, Slieker RC, Luijk R, Dekkers KF, Stein AD, Xu KM, Slagboom PE, van Zwet EW, Lumey LH, Heijmans BT (2018) DNA methylation as a mediator of the association between prenatal adversity and risk factors for metabolic disease in adulthood. Sci Adv 4(1). https://doi.org/10.1126/sciadv.aao4364

von Zglinicki T (2002) Oxidative stress shortens telomeres. Trends Biochem Sci 27(7):339–344. https://doi.org/10.1016/s0968-0004(02)02110-2

Vulliamy T, Marrone A, Goldman F, Dearlove A, Bessler M, Mason PJ, Dokal I (2001) The RNA component of telomerase is mutated in autosomal dominant dyskeratosis congenital. Nature 413(6854):432–435. https://doi.org/10.1038/35096585

Waddington CH (1942) The epigenotype. Endeavour 1:18–20

Waddington CH (1957) The strategy of the genes. A discussion of some aspects of theoretical biology. Allen & Unwin, London

Wamala SP (2001) Large social inequalities behind women's risk of coronary diseases. Unskilled work and family strains are crucial factors. Lakartidningen 98(3):177–181

Waterland RA, Jirtle RL (2003) Transposable elements: targets for early nutritional effects on epigenetic gene regulation. Mol Cell Biol 23(15):5293–5300. https://doi.org/10.1128/MCB.23.15.5293-5300.2003

Waterland RA, Jirtle RL (2004) Early nutrition, epigenetic changes at transposons and imprinted genes, and enhanced susceptibility to adult chronic diseases. Nutrition 20(1):63–68. https://doi.org/10.1016/j.nut.2003.09.011

Waterland RA, Lin J-R, Smith CA, Jirtle RL (2006) Post-weaning diet affects genomic imprinting at the insulin-like growth factor 2 (Igf2) locus. Hum Mol Genet 15(5):705–716. https://doi.org/10.1093/hmg/ddi484

Zhang X, Lin S, Funk WE, Hou L (2013) Environmental and occupational exposure to chemicals and telomere length in human studies. Occup Environ Med 70(10):743–749. https://doi.org/10.1136/oemed-2012-101350

6

Measuring Stress

Psychological stress can result in multiple adverse consequences, resulting not only in a general lack of happiness and success but also in the onset and aggravation of severe medical problems, including psychiatric problems such as **major depression**[1] (Lagraauw et al. 2015; Pariante and Lightman 2008) (see examples in Chaps. 4 and 5). In the face of this connection between chronic stress exposure and severe pathologies, one question became more and more crucial very early in modern stress research: how can stress best be measured? Given the prevalence and multifaceted sources of stress, various methods have been developed and tested to grasp how to quantify stress. In this chapter, we present the most prominent stress measurement methods: targeting acute as well as chronic stress, focusing on minor **stressors** and major crises, building on physiological as well as psychological grounds, and developed for use in laboratories as well as in the field—including everyday real-life conditions as shaped by poverty.

The Basic Challenges of Stress Measurement

Numerous methods have been developed to measure the amount of stress in patients, in humans in general, and in experimental animals. All those investigations are in need of objective physiological and biochemical parameters in order to quantify stress. Such methods have been developed and evaluated over the course of recent decades, with some unexpected difficulties, as we will

---

[1] Glossary terms are bolded at first mention in each chapter.

© Springer Nature Switzerland AG 2021
M. Breitenbach et al., *Stress and Poverty*, https://doi.org/10.1007/978-3-030-77738-8_6

describe below. However, a meaningful correlation between the psychological and the biochemical tests has been found in most cases. It has been highly desirable to develop methods based on reliable biochemical **markers of stress**, and ideally, to compare the psychological and the biochemical measures of stress. Measuring **cortisol** is at the center of stress physiology (despite the measurement difficulties). Cortisol (together with **epinephrine**, also known as **adrenaline**), is a central stress hormone, whose target organs and mechanisms of action are the most well known.

Measuring the degree or amount of stress to which an individual is exposed is quite obviously necessary if we want to find quantitative correlations between the physiology of stress and the social consequences of stress, the stress-induced predisposition to certain unfavorable physiological, mental, and cognitive outcomes, including disease, and their severity. The concept of measuring stress is not a straightforward one, and the different methods for measuring stress that have been developed up to now and are commonly used in stress research require definitions and numerous controls. In the words of George P. Chrousos, one of the leading experts in the field, "stress is ubiquitous and universally pervasive; however, its objective quantification has not been easy" (Chrousos 2009). Accordingly, we will explain in this chapter how stress is usually measured—and why some common methods are preferred to others—as well as why the quantification of stress is not easy and what the most common problems in measuring stress actually are. Importantly for the aim of our book, we will also discuss the degree of agreement between objective measures of stress (measuring biochemical stress markers) and subjective ones (using psychological questionnaires), and what we can learn from these comparisons.

Throughout this book, we are trying to integrate psychological and sociological findings about stress with medical data and biochemical measurements of stress in organisms and particularly in humans, due to our specific interest in chronic poverty as a critical stressor. We think that it is necessary to establish a correlation between the different methods of measuring stress and the results obtained, both by quantifying and evaluating stress markers using analytical biochemical and physiological methods on the one hand and psychological methods on the other. Such correlations are necessary to clarify whether subjectively perceived stress can be related to objectively measured biochemical stress parameters.

To give a quick (yet by no means exhaustive) idea of what the most used measurement methods in these categories are, on the psychological side they include: the **Social Readjustment Rating Scale (SRRS)** established by Holmes and Rahe (1967); the **Perceived Stress Scale (PSS)** developed by

Cohen et al. (1983); and the **Trier Social Stress Test** (**TSST**) introduced by Kirschbaum et al. (1993). On the physiological or biochemical side, examples include the quantitative measurement of cortisol in saliva (cf. Kalman and Grahn 2004) or hair (cf. Slominski et al. 2015); and measurements of heart rate, blood pressure, and related physiological parameters, such as serum epinephrine (cf. Sullivan et al. 2019).

Measuring Acute Stress

Among the biochemical and physiological measures of stress, we must discern between methods for measuring acute stress and methods that measure the long-term effects of stress mediated by the **stress response** system: in other words, the effects of chronic stress. Among the acute measurements, salivary cortisol (Kalman and Grahn 2004; Hellhammer et al. 2009) has been the focus of the large majority of papers. A main reason for this is that the method is easy to apply, non-invasive, and does not immediately require a laboratory setting. The samples can be kept or frozen for some time and can be analyzed later. However, as already indicated, the peculiarity of cortisol secretion and function poses some problems: cortisol shows a diurnal variation with a large peak some time after waking up (Sin et al. 2017) and gradually decreases during the day. Moreover, the baseline value of cortisol depends very much on the individual, making it difficult to define a reference value (Dickerson and Kemeny 2004). Therefore, in cortisol measurements, the **standardized mean change** statistics (Morris 2000; Dickerson and Kemeny 2004) are used as a normalized value for cortisol concentrations. Finally, the relationship between cortisol in serum (which is the value that really matters) to cortisol in saliva (which is more easily measurable) has only been explored in sufficient detail in very few papers (Sin et al. 2017; Steckl and Ray 2018; Ray and Steckl 2019). For practical reasons, though, in the majority of psychological and sociological papers, salivary cortisol has been used as a proxy for the objective acute stress level of probands or patients.

Besides cortisol, epinephrine plays a central role in an organism's stress response system. Epinephrine is an immediate indicator of stress, and it can be measured in serum with an **enzyme immunoassay** (**EIA**) commercial kit (Sullivan et al. 2019). However, in psychological stress research, measurement of epinephrine in serum is a less-used method. A comparison of methods for measuring concentrations of different **biomarkers** of stress with an array of analytical chemical methods in body fluids (serum, sweat, urine, saliva) has been published recently (Steckl and Ray 2018; Ray and Steckl 2019). One example of a surprising result in the 2018 paper is that cortisol concentrations in sweat and saliva differ by a factor of 10, with more cortisol secreted in sweat.

We now want to briefly mention measurements of physiological parameters related to the stress response, which can be of extreme importance for medical diagnosis.

Stress-induced myocardial **ischemia** is measured by intravenous injection of the radioisotope ^{99m}Tc (metastable nuclear isomer of technetium-99) and gamma-imaging of the heart muscle (Hammadah et al. 2018). This method measures an immediate reaction to stress, which leads to a deficit of blood flow in response to the experimentally induced stress in the heart muscle. This can be recognized by gamma-ray imaging using the short-lived radioactive isotope, ^{99m}Tc. This physiological stress reaction is important and highly relevant for patients suffering from **coronary heart disease** and helps to estimate the risk of imminent heart failure.

Leucocyte coping capacity (**LCC**) (McLaren et al. 2003; Huber et al. 2019; Shelton-Rayner et al. 2011) is another method for indirectly measuring the acute reaction to stress. It was originally developed for wildlife biology management due to its ease of use in the field and has since proved to be generally applicable to stress measurement in humans as well. This method is commercially available and therefore more easily reproducible and standardized (MacKenzie 2019). In this method, a small amount of blood is drawn, which is then treated immediately with a micromolar dose of **phorbol 12-myristate 13-acetate** (**PMA**), a pro-carcinogen, and a typical inducer of the macrophage **oxidative burst** contributing to the innate immune reaction. In principle, the method uses the capacity of the blood's macrophages and monocytes to produce superoxide and the **ROS** derived therefrom when stimulated by the pro-carcinogen. This capacity is proportionally reduced under stress. The production of superoxide radicals during the oxidative burst is measured as luminescence units in a portable luminometer. The level of luminescence is inversely proportional to the immediately preceding stress. What the method takes advantage of is a reproducible inhibition of the innate immune reaction in the animal or patient, and it is one of the ways in which the repression of the immune system by stress was shown (Dhabhar 2014). This method has been correlated with other immediate stress parameters, such as blood pressure and heart rate, and was shown to be in excellent agreement (Shelton-Rayner et al. 2011). The results of this test, which can be used easily "in the field," completely corroborate the notion that acute or chronic stress suppresses the immune responses of the stressed person, resulting in absence of, or severely diminished, oxidative burst, an important component of **innate**

immunity exerted by monocytes and macrophages.[2] Further corroborating this finding, we know from medical experience that severely stressed people are more prone to infectious disease than unstressed people (Cohen and Williamson 1991; Øversveen et al. 2017); investigations of the relationship between health and infectious diseases and acute and chronic stress show this as well.

Here we are hinting at stress and stress measurement not merely being limited to physiological and biochemical processes or interventions, but also involving **psychoneuroimmunology** (**PNI**). In a nutshell, PNI describes how, with regard to the immune system, "the ways in which stressors—and the negative emotions they generate—can be translated into physiological changes" (Glaser and Kiecolt-Glaser 2005). In their paper, Glaser and Kiecolt-Glaser actually touch on the topic of poverty, if just briefly, by pointing to "unemployment" as one example of a potential "long-term stressor." This refers to the experience of poverty (as well as other stressors) being powerful both in terms of a single stress-generating event (e.g., losing one's job) and as a long-term or chronic social and psychological stressor.

We will go into more detail below about the measurement of psychosocial stress. Before doing so, we briefly mention further common methods of stress testing that include, for instance, acute exercise stress on a treadmill (Gibbons et al. 2002), which also shows an increase in heart rate, blood pressure, and related physiological parameters. Less frequently, pharmacological induction of stress is used for stress measurement (Beslic et al. 2016), yet given the focus of our book, this approach is not discussed here. Methods as described here, which also serve as an important control in experiments, do not induce psychological stress per se, which typically includes a strong emotional component (e.g., due to unpredictability of stress or of peer group reactions, as in the methods presented below).

Methods of measuring acute physiological stress in humans have usually been combined with inducing social stress or mental stress in laboratory experiments. The reaction to induced social stress has then been correlated with other social or medical parameters, such as low socioeconomic status or a poor neighborhood, a case history of coronary artery disease, or social and medical parameters appearing to be interdependent (Claudel et al. 2018; Sullivan et al. 2019; Soares et al. 2020). The social or mental stress tests which are most often used in the laboratory are variations of the Trier Social Stress

---

[2] See also Chap. 4, where we discuss the influence of stress on the immune system in the examples of psychosomatic diseases.

Test (TSST), which we briefly describe here as one well-established example of this type of measurement.

The TSST was developed by Kirschbaum and colleagues in 1993 in order to overcome some of the limitations faced by experiments using then-common methods to induce psychological stress, e.g., a lack of reliability in regard to stimulations and subsequently to stress responses, hence often leading to inconsistent results (Kirschbaum et al. 1993; Allen et al. 2017; Herhaus and Petrowski 2018). The TSST includes "stress-generating events" such as public speaking in front of an audience and solving a mental arithmetic problem after a short anticipation period. This test has been widely acknowledged, applied in a large variety of experiments, and critically examined as well as developed further with regard to target groups and research topics (Labuschagne et al. 2019; Narvaez Linares et al. 2020; DeJoseph et al. 2019; Miller and Kirschbaum 2019; Zimmer et al. 2019).

Psychological stress tests either measure the stress created in a controlled test situation, such as the TSST just described as well as its related follow-up test designs, or assess the stress experienced in actual everyday life, as in the examples we describe in the sections below. In both cases it is necessary not only to exclude confounding factors but also to carefully select participants, depending on the aim of the study. Only then it is possible to discover differences in stress reactivity depending on physiological, social, or individual factors, and to generate insights in differences in stress behavior: e.g., between different age groups or between different genders (Allen et al. 2017); between normal-weight and obese study participants (Herhaus and Petrowski 2018); or between people from differing socioeconomic backgrounds (Sullivan et al. 2019). Experiments drawing on social stress tests repeatedly showed a marked increase in serum epinephrine (Sullivan et al. 2019) and cortisol (Dickerson and Kemeny 2004), if the stress was (1) uncontrollable, and (2) a threat to the "social self" (for instance because the probands believed that an audience would judge them). It has also been shown that living in a poor environment (e.g., a poor neighborhood), even at an early age, led to a blunted response to the social stress test, which again also correlated with a higher risk of adverse cardiovascular health and behavioral outcomes, including cardiovascular diseases (Raffington et al. 2018; Sullivan et al. 2019).

Investigating Chronic Stress and the Results of a Stressful Life History

Life, at least sometimes, is stressful. Life can be experienced as particularly stressful if a person is facing challenging life events or if she finds herself in adverse living conditions or a disadvantageous environment. Environments and everyday living conditions shaped by poverty, for instance, can very probably be experienced as stressful. They may be experienced as stressful in that they are perceived as being—or in fact are—beyond a person's control, due to issues such as crowding and noise (Ursache et al. 2017), malnutrition and pollution (Bray et al. 2019), or interpersonal conflicts and violence (McEwen 2007; Fagan et al. 2015; Peckins et al. 2020; Finegood et al. 2020).

As we delve into the measurement of stress and at the same time address adverse or poor environments and living conditions, we are shifting our attention to situations that are not adequately referred to solely as incidences of "acute stress" but rather show that the boundaries between one-time, acute stress and cumulative, prolonged, chronic stress can be fluid. Of course, such conditions also include acute stress-generating situations, and maybe even plenty of them. Yet they are—in our core example of poverty and beyond—very likely to involve situations and circumstances characterized by severe strain that cast a lasting cloud over a person's or a family's life or even over that of a whole community. This brings us to the question of how to measure cumulative, prolonged, or chronic stress. To give an idea of the possibilities, we will briefly introduce below some exemplary methods often grouped together as "life event scales" or "checklist measures" (Wethington 2016; for a comprehensive overview see Cohen et al. 1995; Dohrenwend 2006).

One of the most famous approaches to stressful life events is the Social Readjustment Rating Scale (SRRS), developed by Holmes and Rahe (1967), also known as the "Holmes–Rahe Stress Scale" or, particularly highlighting the idea behind it, the "Holmes–Rahe Stress Inventory." The authors built on previous studies, e.g., Adolf Meyer's highly influential studies and Harold G. Wolff and associates' advancements of Meyer's work (Holmes and Rahe 1967; Cohen et al. 1995), that had already linked stressful life events to incidences of illness following a checklist approach. When developing their scale, Holmes and colleagues aimed at deepening the understanding of such causalities (Holmes and Rahe 1967; Rahe et al. 1970). Their 1967 scale originally included 43 items "empirically derived from clinical experience" and has since been adapted time and again, just as the TSST has, in order to reflect possible changes in lifestyle as well as to suit specific target groups such as children and adolescents (Cohen et al. 1983; Dohrenwend 2006; Taylor 2015).

The items listed in the SRRS are each tied to a specific value, reflecting and weighting the effort and time necessary to readjust to a specific event. The listed events include both major and minor occurrences, more or less under a person's control; they range from massive and lasting disruptions, such as the death of a spouse (scoring 100 points) to incidents such as minor violations of the law (11 points), with the commonality of indicating or requiring a significant change in a person's ongoing life pattern (Holmes and Rahe 1967). Values in this scale are set, and while individuals are able to choose which event has actually affected them in a given time frame, at least in the original scale there is no option to weigh events according to their disturbing impact from one's personal point of view and experience. It is noteworthy, though, that the researchers themselves did not set the specific values tied to the life events listed in the SRRS. Rather, the values were empirically assigned during the development phase of the SRRS by nearly 400 participants assessing the relative necessary time and effort for readjustment, hence mirroring a kind of collective perception of the stressfulness of the selected events (Holmes and Rahe 1967).

The values for critical or stressful life events accumulating within a given time frame (e.g., within the year previous to filling out the questionnaire), allow assumptions about the amount of stress a person is suffering due to life changes and, furthermore, allow valid predictions of the near-future health-related impact due to stress (Holmes and Rahe 1967; Rahe et al. 1970).

In a book on "stress and poverty," we need to comment on one detail here: among the items included in the SRRS there are no explicit poverty-related events, but entry points for social decline or even poverty can be found, such as the death of a spouse, divorce, job loss, or the foreclosure of a mortgage (all of these have around 50 points or significantly higher). Remarkably, in the SRRS there are also events listed such as vacation (13 points) or Christmas (12 points); in recent versions, the last item is often generalized to "major holiday." In an SRRS-based study on stress conducted with participants classified as "working poor," i.e., people who are working (often full time) yet earn an income below a given poverty line, "Christmas" turned out to be the event most frequently marked as stressful (by 228 out of 358 respondents in total), followed at some distance by suffering an injury or a severe illness (132 mentions). As an explanation, it seems plausible that, especially in families with children, the mental strain and perceived financial demands needed to enjoy this holiday, which to so many people is very important, far exceed the resources available (Řimnáčová et al. 2018). Possibly surprising examples of "stressful life events," such as of vacations or major holidays, mirror an insight once expressed by Aaron Antonovsky: "what is important for their consequences is the subjective perception of the meaning of the event rather than its objective character" (Antonovsky 1974).

Despite Antonovsky's skepticism about checklist approaches and stress research focusing on the mere occurrence of major or minor life events (Antonovsky 1990), we surely would not want to omit the important contributions of Richard S. Lazarus and colleagues to the understanding of stress, particularly when it comes to poverty-related stress. The most important causative factors that they found are not only related to "major" life events, but also, to supposedly minor "day-to-day events" or "daily hassles," and how they are perceived—"appraised," in the terminology of Lazarus and colleagues (Kanner et al. 1981; Lazarus and Folkman 1984; DeLongis et al. 1988, all co-authored by Lazarus).

In the just-cited studies, based on questionnaires relating to daily hassles (and, respectively, uplifts), which not only ask for frequency but also for perceived severity of events taking place (ranging from measures of 1 to 3, covering "somewhat," "moderately," and "extremely" severe events), the researchers recognize the cumulative impact even of seemingly unimportant, unfavorable happenings. The list, consisting of not less than 117 daily hassles alone, includes, for instance, social obligations, troubling thoughts about the future, various financial issues (or consequences thereof), overstrain and worries about different problems, and health issues or environmental problems such as an evidently deteriorating neighborhood (Kanner et al. 1981). It has to be said that however popular the concept of daily hassles and their measurement has become in stress research, Lazarus and colleagues—and unsurprisingly, given the breadth of their list—have also been heavily criticized for confounding symptoms of health disorders with events or "hassles" as causes (see, for example, Dohrenwend et al. 1984). However, the concept is convincing in that such seemingly "mundane realities of daily life ..., when experienced cumulatively, could be quite stressful" (Wheaton et al. 2013). Given their likely reoccurrence, it seems plausible that even minor events may well qualify not only as single stress-generating events but also as chronic stressors. Stress researcher Blair Wheaton has found a convincing image for the impact of minor, but lasting, daily hassles, referencing the 1983 collapse of the Mianus River Bridge on the major highway between New York and Boston, which resulted in endless questions about what caused it: "there was no 'life event' that triggered the collapse ... there had been no sudden trauma to the bridge. But long-term rusting ... had finally reached the point where the current structure could survive no more. In Neil Young's words, 'Rust never sleeps.'

The bridge had collapsed due to chronic stress" (Wheaton 1997). The stressful, recurring, day-to-day events such as those described as daily hassles by Lazarus and colleagues can be imagined as the "rust that never sleeps," the daily "hassles" caused by circumstances of poverty. We have seen such long-term effects of chronic stress previously in the example of telomere shortening in mothers of chronically ill children (Epel et al. 2004).

To close this chapter and this first part of our book, we will briefly outline further methods of measuring how chronic stressors are perceived (or appraised, in Lazarus's terminology) as well as measurements of their physiological outcomes.

Perceived Psychological Stress and Physiological Stress Markers

Perceived stress has already been mentioned, and it is at the core of some exemplary methods for measuring psychological stress that we will outline below. We will do so in the context of discussing a further set of measurement techniques as well as the biochemical and specific physiological stress markers involved. These markers do not represent immediate or acute stress, but account for long-term or chronic stress and strain experiences, and, subsequently, as a long-term response to chronic stress, are deposited as biomarkers in the body and remain there permanently (or at least for a very long time). All the biomarkers to be discussed here result directly or indirectly from cellular **oxidative stress** (see Chap. 3), which both constitutes, and is a consequence of, chronic psychological stress.

On the psychological side, perceived chronic stress is probably best monitored by the Perceived Stress Scale, commonly referred to as the PSS, first published by Sheldon Cohen and colleagues in 1983 (Cohen et al. 1983; Cohen and Williamson 1988 cited in Epel et al. 2004; Roberti et al. 2006). Critically, the PSS builds on the idea of assessing stressful events and also on the idea of individual, subjective, and resource-dependent appraisal, as proposed by Lazarus and colleagues, and accompanies these ideas with psychometrically valid measures of perceived stress (Cohen et al. 1983). It is a self-reported measure with 14 (or in another version 10) questions about the occurrence and frequency of certain feelings and thoughts with five options for answering (ranging from "never" to "very often"); for instance: "In the last month, how often have you been upset because of something that happened unexpectedly?" or "In the last month, how often have you felt that you were unable to control the important things in your life?" The PSS then uses a point system to evaluate how often in the last month the person felt unable to cope with the stress of life (Cohen et al. 1983). Wheaton and colleagues have argued that in emphasizing perception the PSS actually measures "the outcome of stress, rather than the stressors themselves" (Wheaton et al. 2013). This is anticipated by Cohen and colleagues, who state that a person's "perceived stress can be viewed as an outcome variable—measuring the experience of objective stressful events, coping processes, personality factors, etc." (Cohen et al. 1983). In short, the questions assessing the perception of stress ask about experiences of perceived control, reliability, and security or the lack thereof. Most of them, though not speaking explicitly of poverty, exclusion, or deprivation, could reasonably be related to the experience of poverty.

The Perceived Stress Scale has been widely used and adopted in research to generally assess how stressful one's life is or feels and to study outcomes in both mental and physical health (Cohen 2000; Wheaton et al. 2013). To give another example of a scale meant to assess chronic stress, inspired in part by the PSS, we refer to the **Trier Inventory for the Assessment of Chronic Stress** (**TICS**, Schulz and Schlotz 1999). First created by Peter Schulz and Wolff Schlotz in 1999, the TICS has since received much attention, and—such as its follow-up designs—has been repeatedly and successfully tested for validity (Kromm et al. 2010; Petrowski et al. 2012; Wheaton et al. 2013; Petrowski et al. 2018).

The TICS was developed in order to better understand the impact of chronic psychological stress on health. The focus of the inventory is on "demands placed on individuals in occupational and social settings, as well as imbalances in personal traits and resources" (Petrowski et al. 2018). Similar to the PSS, participants are asked to rate the occurrence and frequency, within 3 months preceding the assessment, of certain situations that may cause overstrain. These include, for instance, experiences of work overload, work discontent, social strain, lack of social recognition, but also anxieties and worries or oppressive memories regarding past or forthcoming events—including "nonevents," i.e., anticipated and desired events or changes that do not occur, leaving a person "in an undesired status" (Wheaton et al. 2013; see also Schulz and Schlotz 1999). Clearly, the TICS touches realms of the mind that could be very likely affected by poverty-related psychological stress.

As various studies using these different methods of stress evaluation show, major disrupting life events, such as the death of a close relative or partner, can cause a permanent disorder (e.g., **post-traumatic stress disorder**) and have to be accounted for. Generally, to define a psychological or psychosocial stress index, three common measures of psychological stress are combined:

"the number of stressful life events, the degree that a participant felt that current demands exceeded his or her ability to cope, and scores from a negative emotion word list (including words such as 'sad', 'angry', or 'nervous')" (Glaser and Kiecolt-Glaser 2005).

We have already discussed the physiological outcomes of acute stress in terms of biomarkers. The same is also true, of course, concerning chronic stress. We will give a few examples for biomarkers related to chronic stress in the following section and conclude this chapter by returning to an impressive example of chronic psychological stress that leads to adverse health consequences.

Biochemical Long-Term Stress Markers

On the biochemical side, we see objective laboratory results related to the experience of stress and stress response. The long-term stress markers concern biochemical changes that are not rapidly metabolized and renewed but instead add up over time, and in some cases allow an estimation of the sum of (oxidative) hits, which the person has experienced. Examples of the measurable results of chronic stress in terms of biomarkers include:

- Measurement of **8-oxo-deoxyguanosine**, a DNA oxidation product, in the DNA of leucocytes (Irie et al. 2001, 2003)
- Measurement of F_2-isoprostanes in urine normalized to creatinine and vitamin E (alpha-tocopherol) as a proxy for lipid peroxidation defining an index for oxidative stress (Epel et al. 2004)
- Measurement of telomere length, which strongly depends on oxidative stress and **telomerase** activity (Epel et al. 2004), as described in Chap. 5
- Measurement of inflammatory markers; although acute stress weakens the respiratory burst of macrophages and monocytes in the innate immune system and although long-term stress can suppress both the innate and the adaptive immune system (Dhabhar 2014), certain components of the inflammatory cascade are strongly induced by stress. Preeminent among the components measured is **IL-6**, a cytokine that according to circumstances can be pro-inflammatory or anti-inflammatory. Chronic stress seems to imply a state of low-grade permanent inflammation (Dhabhar 2014).

Perceived Stress and Objective Stress

Can "perceived stress" and "objective stress" be directly compared? Which one is the more reliable risk factor for psychosomatic diseases, diseases of aging, and other unfavorable health-related, mental, or cognitive outcomes? As could be expected, perceived social stress is a powerful disrupting force for individuals, and correlates very well with the physical measurements listed above. This is shown in a number of studies that combine psychological stress measures with physiological stress markers.

As an example, we refer once more to the pioneer paper by Epel et al. (2004), in which the results of perceived stress scale tests, assessing the chronic stress exposure of mothers caring for chronically ill children, were compared to the oxidative stress index defined above, and an excellent correlation was found (see Chap. 5). The findings were additionally corroborated by the fact that attrition of **telomeres** can be caused by oxidative stress. The main, important result of the paper was that mothers caring for a chronically ill child showed short telomeres in their **peripheral blood mononuclear cells** (**PBMC**), a critical physiological parameter related to aging (and hence lifespan) as well as to various health outcomes, including cardiovascular diseases and certain types of cancers (and the risk of dying from certain types of cancer), as discussed in Chap. 4. Telomere shortening was found to be a linear process, depending on the time the mothers spent caring for their sick children. Taken together, perceived stress seemed to be the determinative factor causing physiological stress in the women and, ultimately, the shortening of their telomeres.

Perceived psychological stress and measurable physiological stress are clearly linked, with a broad variety of adverse health outcomes. As we have said before, the experience of poverty raises the likelihood of experiencing or perceiving stress—in single, acute stress-generating events as well as in situations that are cumulative, prolonged, and chronically stressful. This connection between the experiences of poverty and stress is due, generally, to physiological, environmental, and psychological (psychosocial) factors (Cohen et al. 1995), and in particular, an increase in threats, demands, structural constraints, underreward, complexity of everyday life, uncertainty, conflict, restriction of choice, and of course resource deprivation—to cite the categories emphasized by Blair Wheaton in the context of chronic stress (1997). Although stress, in its life-saving as well as life-threatening facets, is obviously to a certain degree part of life in general, we have to bear in mind that health, in its own right, is already not shared equally among human beings. Nor is stress exposure. There is evidence that living conditions shaped by poverty give rise to both general health problems and stress exposure (quite often intertwined) or the outcomes thereof.

In the following chapters, we will focus on stress and poverty from the perspective of poverty. We will discuss research findings that link poverty or low socioeconomic status with stress and hence with health outcomes (often referring to stress measures described above), clarify what makes "poverty" such a **toxic stressor**, and suggest how adverse outcomes of poverty-related stress might be avoided or at least cushioned.

References

Allen AP, Kennedy PJ, Dockray S, Cryan JF, Dinan TG, Clarke G (2017) The Trier Social Stress Test: principles and practice. Neurobiol Stress 6:113–126. https://doi.org/10.1016/j.ynstr.2016.11.001

Antonovsky A (1974) Conceptual and methodological problems in the study of resistance resources and stressful life events. In: Dohrenwend BS, Dohrenwend BP (eds) Stressful life events: their nature and effects. Wiley, New York, pp 245–273

Antonovsky A (1990) A somewhat personal odyssey in studying the stress process. Stress Med 6:71–80. https://doi.org/10.1002/smi.2460060203

Beslic N, Milardovic R, Sadija A, Ceric S, Raic Z (2016) Regadenoson in myocardial perfusion study—first institutional experiences in Bosnia and Herzegovina. Acta Informatica Medica 24(4):405–408. https://doi.org/10.5455/aim.2016.24.405-408

Bray R, De Laat M, Godinot X, Ugarte A, Walker R (2019) The hidden dimensions of poverty. International participatory research. Fourth World, Montreuil

Chrousos GP (2009) Stress and disorders of the stress system. Nat Rev Endocrinol 5:374–381. https://doi.org/10.1038/nrendo.2009.106

Claudel SE, Adu-Brimpong J, Banks A, Ayers C, Albert MA, Das SR, de Lemos JA, Leonard T, Neeland IJ, Rivers JP, Powell-Wiley TM (2018) Association between neighborhood-level socioeconomic deprivation and incident hypertension: a longitudinal analysis of data from the Dallas Heart Study. Am Heart J 204:109–118. https://doi.org/10.1016/j.ahj.2018.07.005

Cohen S (2000) Measures of psychological stress. Summary prepared by Sheldon Cohen with the Psychosocial Working Group. Last revised February, 2000. https://macses.ucsf.edu/research/psychosocial/stress.php. Accessed 26 Jan 2021

Cohen S, Williamson GM (1988) Perceived stress in a probability sample of the United States. In: Spacapan S, Oskamp S (eds) The social psychology of health. Sage, Newbury Park, CA, pp 31–67

Cohen S, Williamson GM (1991) Stress and infectious disease in humans. Psychol Bull 109(1):5–24. https://doi.org/10.1037/0033-2909.109.1.5

Cohen S, Kamarck T, Mermelstein R (1983) A global measure of perceived stress. J Health Soc Behav 24(4):385–396. https://doi.org/10.2307/2136404

Cohen S, Kessler RC, Underwood Gordon L (1995) Measuring stress. A guide for health and social scientist. Oxford University Press, New York

DeJoseph ML, Finegood ED, Raver CC, Blair CB, The Family Life Project Key Investigators (2019) Measuring stress reactivity in the home: preliminary findings from a version of the Trier Social Stress Test (TSST-H) appropriate for field-based research. https://doi.org/10.31234/osf.io/5qapw

DeLongis A, Folkman S, Lazarus RS (1988) The impact of daily stress in health and mood: psychological and social resources as mediators. J Pers Soc Psychol 54(3):486–495. https://doi.org/10.1037//0022-3514.54.3.486

Dhabhar FS (2014) Effects of stress on immune function: the good, the bad, and the beautiful. Immunol Res 58:193–210. https://doi.org/10.1007/s12026-014-8517-0

Dickerson SS, Kemeny ME (2004) Acute stressors and cortisol responses: a theoretical integration and synthesis of laboratory research. Psychol Bull 130(3):355–391. https://doi.org/10.1037/0033-2909.130.3.355

Dohrenwend BP (2006) Inventorying stressful life events as risk factors for psychopathology: toward resolution of the problem of intracategory variability. Psychol Bull J 132(3):477–495. https://doi.org/10.1037/0033-2909.132.3.477

Dohrenwend BS, Dohrenwend BP, Dodson M, Shrout PE (1984) Symptoms, hassles, social supports, and life events: problem of confounded measures. J Abnorm Psychol 93(2):222–230. https://doi.org/10.1037//0021-843x.93.2.222

Epel ES, Blackburn EH, Lin J, Dhabhar FS, Adler NE, Morrow JD, Cawthon RM (2004) Accelerated telomere shortening in response to life stress. Proc Natl Acad Sci U S A 101(49):17312–17315. https://doi.org/10.1073/pnas.0407162101

Fagan AA, Wright EM, Pinchevsky GM (2015) Exposure to violence, substance use, and neighborhood context. Soc Sci Res 49:314–326. https://doi.org/10.1016/j.ssresearch.2014.08.015

Finegood ED, Chen E, Kish J, Vause K, Leigh AKK, Hoffer L, Miller GE (2020) Community violence and cellular and cytokine indicators of inflammation in adolescents. Psychoneuroendocrinology 115:1–10. https://doi.org/10.1016/j.psyneuen.2020.104628

Gibbons RJ, Balady GJ, Bricker JT, Chaitman BR, Fletcher GF, Froelicher VF, Mark DB, McCallister BD, Mooss AN, O'Reilly MG, Winter WL Jr (2002) ACC/AHA 2002 guideline update for exercise testing: summary article. A report of the American College of Cardiology/American Heart Association Task Force on Practice Guidelines (Committee to update the 1997 exercise testing guidelines). J Am Coll Cardiol 40(8):1531–1540. https://doi.org/10.1016/s0735-1097(02)02164-2

Glaser R, Kiecolt-Glaser JK (2005) Stress-induced immune dysfunction: implications for health. Nat Rev Immunol 5:243–251. https://doi.org/10.1038/nri1571

Hammadah M, Sullivan S, Pearce B, Al Mheid I, Wilmot K, Ramadan R, Tahhan AS, O'Neal WT, Obideen M, Alkhoder A, Abdelhadi N, Kelli HM, Ghafeer MM,

Pimple P, Sandesara P, Shah AJ, Hosny KM, Ward L, Ko Y-A, Sun YV, Weng L, Kutner M, Bremner JD, Sheps DS, Esteves F, Raggi P, Vaccarino V, Quyyumi AA (2018) Inflammatory response to mental stress and mental stress induced myocardial ischemia. Brain Behav Immun 68:90–97. https://doi.org/10.1016/j.bbi.2017.10.004

Hellhammer DH, Wüst S, Kudielka BM (2009) Salivary cortisol as a biomarker in stress research. Psychoneuroendocrinology 34:163–171. https://doi.org/10.1016/j.psyneuen.2008.10.026

Herhaus B, Petrowski K (2018) Cortisol stress reactivity to the Trier Social Stress Test in obese adults. Obes Facts 11:491–500. https://doi.org/10.1159/000493533

Holmes TH, Rahe RH (1967) The social readjustment rating scale. J Psychosom Res 11:213–218. https://doi.org/10.1016/0022-3999(67)90010-4

Huber N, Marasco V, Painer J, Vetter SG, Göritz F, Kaczensky P, Walzer C (2019) Leukocyte coping capacity: an integrative parameter for wildlife welfare within conservation interventions. Front Vet Sci 6. https://doi.org/10.3389/fvets.2019.00105

Irie M, Asami S, Nagata S, Miyata M, Kasai H (2001) Relationships between perceived workload, stress and oxidative DNA damage. Int Arch Occup Environ Health 74:153–157. https://doi.org/10.1007/s004200000209

Irie M, Asami S, Ikeda M, Kasai H (2003) Depressive state relates to female oxidative DNA damage via neutrophil activation. Biochem Biophys Res Commun 311:1014–1018. https://doi.org/10.1016/j.bbrc.2003.10.105

Kalman BA, Grahn RE (2004) Measuring salivary cortisol in the behavioral neuroscience laboratory. J Undergrad Neurosci Educ 2(2):A41–A49

Kanner AD, Coyne JC, Schaefer C, Lazarus RS (1981) Comparison of two modes of stress measurement: daily hassles and uplifts versus major life events. J Behav Med 4(1):1–39. https://doi.org/10.1007/BF00844845

Kirschbaum C, Pirke K-M, Hellhammer DH (1993) The 'Trier Social Stress Test'—a tool for investigating psychobiological stress responses in a laboratory setting. Neuropsychobiology 28:76–81. https://doi.org/10.1159/000119004

Kromm W, Gadinger MC, Schneider S (2010) Peer ratings of chronic stress: can spouses and friends provide reliable and valid assessments of a target person's level of chronic stress? Stress Health 26:292–303. https://doi.org/10.1002/smi.1297

Labuschagne I, Grace C, Rendell P, Terrett G, Heinrichs M (2019) An introductory guide to conducting the Trier Social Stress Test. Neurosci Biobehav Rev 107:686–695. https://doi.org/10.1016/j.neubiorev.2019.09.032

Lagraauw HM, Kuiper J, Bot I (2015) Acute and chronic psychological stress as risk factors for cardiovascular disease: insights gained from epidemiological, clinical and experimental studies. Brain Behav Immun 50:18–30. https://doi.org/10.1016/j.bbi.2015.08.007

Lazarus RS, Folkman S (1984) Stress, appraisal, and coping. Springer, New York

MacKenzie RJ (2019) A new test claims to be able to measure stress in a drop of blood. https://www.technologynetworks.com/neuroscience/articles/can-we-really-measure-stress-320611. Accessed 26 Jan 2021

McEwen BS (2007) Physiology and neurobiology of stress and adaptation: central role of the brain. Physiol Rev 87:387–904. https://doi.org/10.1152/physrev.00041.2006

McLaren GW, Macdonald DW, Georgiou C, Mathews F, Newman C, Mian R (2003) Leukocyte coping capacity: a novel technique for measuring the stress response in vertebrates. Physiol Soc 88(4):541–546. https://doi.org/10.1113/eph8802571

Miller R, Kirschbaum C (2019) Cultures under stress: a cross-national meta-analysis of cortisol responses to the Trier Social Stress Test and their association with anxiety-related value orientations and internalizing mental disorders. Psychoneuroendocrinology 105:147–154. https://doi.org/10.1016/j.psyneuen.2018.12.236

Morris SB (2000) Distribution of the standardized mean change effect size for meta-analysis on repeated measures. Br J Math Stat Psychol 53:17–29. https://doi.org/10.1348/000711000159150

Narvaez Linares NF, Charron V, Ouimet AJ, Labelle PR, Plamondon H (2020) A systematic review of the Trier Social Stress Test methodology: issues in promoting study comparison and replicable research. Neurobiol Stress 13:10235. https://doi.org/10.1016/j.ynstr.2020.100235

Øversveen E, Rydland HT, Bambra C, Eikemo TA (2017) Rethinking the relationship between socio-economic status and health: making the case for sociological theory in health inequality research. Scand J Public Health 45:103–112. https://doi.org/10.1177/1403494816686711

Pariante CM, Lightman SL (2008) The HPA axis in major depression: classical theories and new developments. Trends Neurosci 31(9):464–468. https://doi.org/10.1016/j.tins.2008.06.006

Peckins MK, Roberts AG, Hein TC, Hyde LW, Mitchell C, Brooks-Gunn J, McLanahan SS, Monk CS, Lopez-Duran NL (2020) Violence exposure and social deprivation is associated with cortisol reactivity in urban adolescents. Psychoneuroendocrinology 111:104426. https://doi.org/10.1016/j.psyneuen.2019.104426

Petrowski K, Paul S, Albani C, Brähler E (2012) Factor structure and psychometric properties of the trier inventory for chronic stress (TICS) in a representative German sample. BMC Med Res Methodol 12:42. https://doi.org/10.1186/1471-2288-12-42

Petrowski K, Kliem S, Sadler M, Meuret AE, Ritz T, Brähler E (2018) Factor structure and psychometric properties of the English version of the trier inventory for chronic stress (TICS-E). BMC Med Res Methodol 18:18. https://doi.org/10.1186/s12874-018-0471-4

Raffington L, Prindle J, Keresztes A, Binder J, Heim C, Shing YL (2018) Blunted cortisol stress reactivity in low-income children relates to lower memory function.

Psychoneuroendocrinology 90:110–121. https://doi.org/10.1016/j.psyneuen.2018.02.002

Rahe RH, Mahan JL, Arthur RJ (1970) Prediction of near-future health change from subjects' preceding life changes. J Psychosom Res 14:401–406. https://doi.org/10.1016/0022-3999(70)90008-5

Ray P, Steckl AJ (2019) Label-free optical detection of multiple biomarkers in sweat, plasma, urine, and saliva. ACS Sensors 4:1346–1357. https://doi.org/10.1021/acssensors.9b00301

Řimnáčová Z, Kajanová A, Ondrášek S (2018) The working poor and stress. J Nurs Soc Stud Public Health Rehabil 3–4:93–100

Roberti JW, Harrington LN, Storch EA (2006) Further psychometric support for the 10-item version of the perceived stress scale. J Coll Couns 9:135–147. https://doi.org/10.1002/j.2161-1882.2006.tb00100.x

Schulz P, Schlotz W (1999) Trierer Inventar zur Erfassung von chronischem Streß (TICS): Skalenkonstruktion, teststatistische Überprüfung und Validierung der Skala Arbeitsüberlastung. Diagnostica 45(1):8–19. https://doi.org/10.1026//0012-1924.45.1.8

Shelton-Rayner GK, Mian R, Chandler S, Robertson D, Macdonald DW (2011) Quantifying transient psychological stress using a novel technique: changes to PMA-induced leukocyte production of ROS in vitro. Int J Occup Saf Ergon 17(1-3):3–13. https://doi.org/10.1080/10803548.2011.11076866

Sin NL, Ong AO, Stawski RS, Almeida DM (2017) Daily positive events and diurnal cortisol rhythms: examination of between-person differences and within-person variation. Psychoneuroendocrinology 83:91–100. https://doi.org/10.1016/j.psyneuen.2017.06.001

Slominski R, Rovnaghi CR, Anand KJS (2015) Methodological considerations for hair cortisol measurements in children. Ther Drug Monit 37(6):812–820. https://doi.org/10.1097/FTD.0000000000000209

Soares S, Santos AC, Soares Peres F, Barros H, Fraga S (2020) Early life socioeconomic circumstances and cardiometabolic health in childhood: evidence from the generation XXI cohort. Prev Med 133:106002. https://doi.org/10.1016/j.ypmed.2020.106002

Steckl AJ, Ray P (2018) Stress biomarkers in biological fluids and their point-of-use detection. ACS Sensors 3:2025–2044. https://doi.org/10.1021/acssensors.8b00726

Sullivan S, Kelli HM, Hammadah M, Topel M, Wilmot K, Ramadan R, Pearce BD, Shah A, Lima BB, Kim JH, Hardy S, Levantsevych O, Obideen M, Kaseer B, Ward L, Kutner M, Hankus A, Ko Y-A, Kramer MR, Lewis TT, Brammer JD, Quyyumi A, Vaccarino V (2019) Neighborhood poverty and hemodynamic, neuroendocrine, and immune response to acute stress among patients with coronary artery disease. Psychoneuroendocrinology 100:145–155. https://doi.org/10.1016/j.psyneuen.2018.09.040

Taylor JM (2015) Psychometric analysis of the ten-item Perceived Stress Scale. Psychol Assess 27(1):90–101. https://doi.org/10.1037/a0038100

Ursache A, Merz EC, Melvin S, Meyer J, Noble KG (2017) Socioeconomic status, hair cortisol and internalizing symptoms in parents and children. Psychoneuroendocrinology 78:142–150. https://doi.org/10.1016/j.psyneuen.2017.01.020

Wethington E (2016) Live events scale. In: Fink G (ed) Stress: concepts, cognition, emotion, and behavior, Handbook of stress series, vol 1. Academic, London, pp 103–108. https://doi.org/10.1016/B978-0-12-800951-2.00012-1

Wheaton B (1997) The nature of chronic stress. In: Gottlieb BH (ed) Coping with chronic stress. Plenum, New York, pp 43–73. https://doi.org/10.1007/978-1-4757-9862-3_2

Wheaton B, Young M, Montazer S, Stuart-Lahman K (2013) Chapter 15: Social stress in the twenty-first century. In: Aneshensel CS, Phelan JC, Bierman A (eds) Handbook of the sociology of mental health, 2nd edn. Springer, Amsterdam, pp 299–323. https://doi.org/10.1007/978-94-007-4276-5_15

Zimmer P, Buttlar B, Halbeisen G, Walther E, Domes G (2019) Virtually stressed? A refined virtual reality adaptation of the Trier Social Stress Test (TSST) induces robust endocrine responses. Psychoneuroendocrinology 101:186–192. https://doi.org/10.1016/j.psyneuen.2018.11.010

7

The Language Games of Stress

In Chaps. 2–6, we have described how stress research evolved from the study of single cells (e.g., yeast) to complex organisms, and then again from animals (e.g., rodents) to human beings. We have discussed the manifold and often detrimental and lasting consequences that excessive and chronic stress exposure can bring about. By looking at different methods of stress measurement that were further and further elaborated and refined during decades of research, we have seen that stress not only becomes visible through physiological **markers**[1]—as objective, measurable stress—but that stress also has a psychological side. We can find objective markers and subjective evidence as well. Both dimensions have to be considered. The subjective aspect is approached with terms such as "perceived stress" or "stress experience." There is not just one way to think about stress and to speak about stress. Consequently, let us unpack the **language games** of stress.

Language games are definable contexts of language usage; they are characterized by rules and show complex connections between linguistic aspects, such as words and sentences, and extra-linguistic aspects, such as gestures and certain types of behavior. The concept of language games, as developed in Ludwig Wittgenstein's *Philosophical Investigations* (Wittgenstein 1967), points to a theory of meaning that sees the meaning of a word inextricably linked with the way the word is being used. The reality of language usage cannot be replaced by exact definitions in specialized circles. There is a certain democracy to the use of language that cannot be controlled by experts.

---

[1] Glossary terms are bolded at first mention in each chapter.

© Springer Nature Switzerland AG 2021
M. Breitenbach et al., *Stress and Poverty*, https://doi.org/10.1007/978-3-030-77738-8_7

If we are to reflect on the landscape of the language game(s) of stress we cannot ignore the complexity that goes beyond a consistent or one-dimensional understanding of the word. This book is a witness to such complexity, by bringing together at least two discourses, the biochemical and neurophysiological discourse on stress and the discourse on stress in the social sciences and especially in poverty research. The book works with a concept ("stress") that is widely used. Given Wittgenstein's approach to the meaning of a word based on its use, we could suspect that each significant use of the term "stress" adds semantic layers, nuances, and connotations. The result is that the term "stress" is not owned by a single discipline, is not housed in a dominant discourse, is not controlled by clearly defined semantic gate-keepers: "The reference to stress is ubiquitous in modern society, and the term *stressful* is a recurrent descriptor of negative experiences related to anything from daily hassles, relationship issues, and pressures at work to health concerns and debilitating phobias. It is interesting that most popular definitions would likely describe purely psychological phenomena, yet less than 100 years ago, the term *stress* as a psychological phenomenon did not exist. Today the concept of stress is pervasive in popular as well academic literature. Despite its prevalence, stress remains an elusive concept" (Robinson 2018).

This elusiveness of the term is both a semantic risk, namely vagueness, and a linguistic opportunity to bridge the gap between disciplines—in the case of our book by bringing biochemistry and the social sciences into dialogue with each other. The wide use of the term does not mean that we cannot find reference points for rules governing the use of the word "stress." There is actually a lot that we can say.

Stress is a noun (defined, according to the Cambridge Dictionary, for example, as "great worry caused by a difficult situation, or something that causes this condition") and also a verb ("to be or feel worried and nervous"). An experience, like working as a frontline worker in a pandemic, can be "stressful," i.e., taxing, characterized by pressure. Candidates for synonyms for "stress" could be the terms "pressure," "strain," "tension," "exertion," "struggle," and "burden."

The word "stress" is used in a number of ways to express mostly adverse conditions. Stress can be, so the language usage goes, "combatted," "managed," "relieved," "reduced," "created." In this sense, stress is a phenomenon that can be shaped and reacted to.

"Stress" points to something that cannot easily be ignored, which is also highlighted by the meaning of "stress" in linguistics as "emphasis," as the way that a word is pronounced with greater force than other words, or as a term that expresses the importance of the weight of something ("X stressed the importance of A"). If the word is used in this sense, candidates for synonyms could be "emphasis," "significance," "weight," and "value."

"Stress" can itself be seen as a force that can change the shape and other characteristics of an object. Stress "does something" to an object. A person can be "under stress," underlining the idea of stress as a force that puts pressure on a person. Stress can "drive a person to do something" or "drive a person to the brink of a nervous breakdown"; here again there is the linguistic representation of the idea that stress causes something and also brings about effects. Stress is thus seen both as cause and as effect; stress can be caused, e.g., by working conditions, and it is causally relevant by causing, e.g., unhappiness or health challenges.

"Stress" is either a force or the result of forces, and this force or result of forces both stimulates and constrains the subject of stress. The subject is forced to respond to stress and is shaped by stress, formed and deformed. Stress is a process, but also the result of a process, a state. It could make sense to distinguish between a subject of stress, a context of stress, types of stress, and the effects of stress: X is put under stress of type Y in context C with the effect E.

X can be an organism or a human person, a financial institution, or some kind of material; Y refers to the properties and types of stress (e.g., chronic, emotional, severe); C can be the environment (neighborhood, workplace, cellular level); E can refer to physiological symptoms like exhaustion or to behavioral effects like creativity.

The semantic field of the term is complex since it has traveled across centuries, languages, and disciplines: "It represents an unfortunate transfer of a word used in mechanics to medicine. Coming from old French language, *estresse* (substantive), meaning strain or straining force, and also from a verb, *estrecer*, to put pressure or strain on something or a body in order to strain it or deform its shape, it supposes an intensity and a duration of the strain but also an opposite reaction resisting the force. It has been extended to physical or mental tensions or strains and the pressures causing them" (Le Moal 2007).

As indicated above, we use the word "stress" in a way that shows stress can have different characteristics depending on duration ("acute," "chronic," "constant," "repetitive"), on extent ("excessive," "enormous," "considerable," "severe"), on effects ("positive," "negative," or even "toxic"), and on the type ("emotional stress" or "social stress" or "occupational stress" or "parenting stress" or "environmental stress" or "financial stress"). The type of stress also points to the area that is most affected by stress or to the area that causes stress. The word is pervasive and used to cover many different challenges, from mild workplace pressure to unbearable situations of illness or grief.

The term negotiates two dimensions—the dimension of the "inner life" of a person and the dimension of external forces. The history of stress research reflects this two-dimensionality: Claude Bernard, in his 1865 treatise "Introduction à l'étude de la medicine expérimentale," discussed the need for an organism to maintain a constant fluid environment bathing the cells of the body; he used the term *milieu intérieur* to describe this aspect of an organism (Goldstein and Kopin 2007). Walter Cannon then developed the term "**homeostasis**," referring to the dynamic inner equilibrium based on an appropriate response to external factors such as changes in the environment (Goldstein and Kopin 2007). Hans Selye, already mentioned in our book, used the term "stress" in the sense of "the nonspecific response of the body to any demand upon it" (Selye 1976), thus making stress the link between the internal dynamics of an organism and the external forces affecting the organism. The term "stress," according to this semantic history, refers to a relationship, to two factors or "worlds." That is why concepts like equilibrium, balance, or stable order make sense when approaching the meaning of "stress."

One word about "equilibrium," though: "stress" is based both on an equilibrium, and on the loss and lack thereof. The normal ground state is not at thermodynamic equilibrium (consider, for instance, the Na+ and K+ concentrations inside and outside the cell). Rather, a certain nonequilibrium must be maintained to enable the living cell to answer to internal as well as external challenges. When thermodynamic equilibrium is reached, the cell is dead. Homeostasis means that this non-equilibrium is not absolutely fixed but is retained at an average value with continuous fluctuation around this average value. Under stress (see Fig. 12.1 in Chap. 12) a new and different homeostatic value of the electrochemical potential gradient is attained. This new "set point" is still compatible with life, but cellular metabolism is different from the ground state. Hence, the term "equilibrium" that has been used to approach the term "stress" cannot be understood as a fixed order. Stress is the dynamic expression of dynamic developments.

There is a "line" where the term "stress" seems inadequate and is replaced by other terms. This line is constituted by the enormity of the adversity, as observed by Ruth Keil: "it is interesting to note cases where the term would not be used: it would sound odd, in a discussion of modern wars and ethnic conflicts, for instance, to say that ethnic cleansing caused the victims to feel stress. The term no longer works as a metaphor, and the external events are so extreme that they explain themselves without reference to the internal state of the person. There is no longer any need for a term that mediates between external events and the internal state of the individual" (Keil 2004). "Stress" makes sense when both dimensions, the internal and the external, are reasonably relevant (a dead body will not suffer stress) and also appropriately moderate (extreme external conditions, as Keil notes, do not support the use of the term "stress" according to our implicit linguistic conventions).

The word "stress" has an increasingly complex history, and the same applies to the history of stress research (Szabo et al. 2017). The concept traveled from medicine and physiology to psychology and from there into ordinary language, but also into the social sciences. According to the Merriam-Webster Dictionary, it seems that the term made a first appearance in the fourteenth century. The etymology of the word "stress" takes us to the Latin "*distringere*" ("to stretch out"), where something is stretched because of an external force; this also means that something is "tested" by experimenting with its adaptability and formability. There is a moment of violence, of aggression that changes the object of the stress. The semantic field of "*distringere*" (vulgar Latin "*districtia*") points to "narrowness," and "narrowness" metaphorically understood points to "affliction," a meaning taken up by the Old French word "*destresse*." Something is pressed together or drawn tight. It is interesting that the term refers both to compression and to stretching out. This inconsistency can be remedied by looking at the key idea of stress in the physical world: "In the physical world, stress can be seen as the progress towards structural failure: it can only be measured by reference to its effects on a structure. If materials never deformed or failed, there would be nothing to observe, and stress would not measure anything. Therefore, it seems reasonable to suggest that what we generally mean by stress, in the physical world, is a measure of the tendency of a force to cause damage or change of state to a structure" (Keil 2004).

The description of stress from a biochemical perspective is quite close to the description taken from the physical world; as a reminder, here is Lushchak's definition of stress, which we used in the introduction: "**Oxidative stress** is a situation when steady state **ROS (reactive oxygen species)** concentration is transiently or chronically enhanced, disturbing cellular metabolism and its regulation and damaging cellular constituents" (Lushchak 2014). The language of "disturbing" and "damaging" is noteworthy.

Because of the semantic connection between "stress" and "damage," it is not surprising that the everyday use of the word "stress" is predominantly negative. Stress is an adverse condition; it is unwanted and something that calls for coping and responding. Stress suggests a threat to the equilibrium, an idea we already find in Cannon's work, and the underlying assumption is that there is a connection between "equilibrium" and the "functioning" of an organism: a certain level of stress may be necessary for the organism to function, but this level must not transcend a threshold where the organism runs out of coping resources and responsive options.

Let us conclude these observations on the language game(s) of "stress" by accepting three key messages: (1) the term "stress" refers to a relationship between the internal structure of an organism and the external world; (2) this relationship negotiates "equilibrium," "order," and "stability," but also "change" and "vitality," so that stress is connected in a complex way to the proper functioning of an organism; (3) the term "stress" is part of a semantic field that includes "coping" and terms to express the appropriate responses to stress.

The main idea of this book is that the ways in which "stress" is used in biochemistry and in the social sciences, respectively, are related by way of "family resemblances" (Wittgenstein 1967). This term expresses the idea that we can use a term across different contexts of language usage without having to commit to a clearly identifiable common and constant property. The three observations just mentioned will serve as sufficient grounds to argue that the term "stress," as used in biochemistry and the social sciences, has sufficient "similarity" and "resemblance" across the contexts to justify the use of this one term—even though we will admit that the language game of "stress" in biochemistry and the language game of "stress" in the social sciences are different and distinct language games. In fact, the similarity goes deeper. The same kind of oxidative stress as defined by a change in the electrochemical potential difference across the plasma membrane of a yeast cell is also found in neurons of the **hippocampus** under stress. We could say that the oxidative stress found causes psychological stress in the hippocampus and other brain areas and therefore in the mind of a stressed person.

References

Goldstein DS, Kopin IJ (2007) Evolution of concepts of stress. Stress 10(2):109–120. https://doi.org/10.1080/10253890701288935

Keil RMK (2004) Coping and stress: a conceptual analysis. J Adv Nurs 45(6):659–665

Le Moal M (2007) Historical approach and evolution of the stress concept: a personal account. Psychoneuroendocrinology 32:S3–S9. https://doi.org/10.1016/j.psyneuen.2007.03.019

Lushchak VI (2014) Free radicals, reactive oxygen species, oxidative stress and its classification. Chemico-Biol Interact 224:164–175. https://doi.org/10.1016/j.cbi.2014.10.016

Robinson AM (2018) Let's talk about stress: history of stress research. Rev Gen Psychol 22(3):334–342. https://doi.org/10.1037/gpr0000137

Selye H (1976) Stress without distress. In: Serban G (ed) Psychopathology of human adaptation. Springer, New York, pp 137–146

Szabo S, Yoshida M, Filakovszky J, Juhasz G (2017) "Stress" is 80 years old: from Hans Selye original paper in 1936 to recent advances in GI ulceration. Curr Pharm Des 23(27):4029–4041. https://doi.org/10.2174/1381612823666170622110046

Wittgenstein L (1967) Philosophical investigations. Blackwell, Oxford

8

The Unhealthy Relationship Between Stress and Poverty

In the previous chapter, we described the **language games**[1] that govern the usage of the term "stress." Important aspects of "stress" as a word in use emerged, such as the interplay between an objective and a subjective dimension. We believe that many of the aspects mentioned in the previous chapter can also be applied to the concept of "poverty." Also, we are convinced that, across the disciplines of biology, stress research, and poverty studies, we are discussing similar phenomena when we speak about stress, including the stressful experience of poverty. In the next part of the book, we investigate the relationship between stress and poverty from various angles; we start in this chapter by first reconstructing how poverty emerged as a topic of interest for stress research, drawing on developments in stress research and stress measurement as described in previous chapters. We then argue that poverty has to be understood as a cause of "**toxic stress**." Additionally, we look at the vicious circle of poverty and poverty-related stress, which leads to more adversity and more stress. Finally, we present examples for adverse health and development outcomes identified in stress research that emphasize the detrimental nature of the relationship between stress and poverty.

---

[1] Glossary terms are bolded at first mention in each chapter.

© Springer Nature Switzerland AG 2021
M. Breitenbach et al., *Stress and Poverty*, https://doi.org/10.1007/978-3-030-77738-8_8

From the Language Games of Stress to the Experience of Poverty

Poverty, just like stress, is a word, but also an experience; it is a term, but also a condition, in that poverty refers to adverse living conditions. As an involuntary situation (as opposed to the voluntary poverty we may find in a number of faith traditions), poverty implies the desire to change existing conditions or limitations in obtaining this condition; it refers to a desired "ground state" of a person, and the gap between desires and fulfillment. Given its involuntary nature in our societies, poverty implies damage and violence, a condition that is forced or imposed upon a person.

Poverty can be a cause for adverse conditions, or it can be the consequence, the result of adverse events; often, these two dynamics go hand in hand, forming a vicious circle. X (e.g., discrimination, a health issue, unemployment) leads to poverty, which causes further X (e.g., discrimination, a health issue, unemployment).

Conditions of poverty cannot be properly understood by analyzing the level of the individual or the household. Even though poverty may isolate people and lead to **social exclusion**, the experience of poverty is not an isolated one. Poverty has to be understood in terms of the connections between individual conditions and internal dynamics within an individual (a family, a neighborhood, or community) on the one hand, and external (social, political) forces affecting the individual (the family, the community) on the other. Conditions of poverty constitute an unwanted, uncontrollable, and often inescapable test of a person's (or a family's or a community's) adaptability and coping skills. Being poor is stressful; poverty points to disruptions in the successful interactions between the individual and the environment, between a social system and its context. In fact, the way individuals and social systems respond to poverty can reveal patterns of **resilience** (see Chap. 12).

Because of the major influence of the socially and politically structured environment, poverty is not simply "happening"; poverty is caused. As an involuntary condition, poverty is frequently imposed on individuals and families, as a result of social dynamics, economic mechanisms, and political decisions. There are undeniable moral and political aspects of poverty that call for an appropriate response.

The following chapters of this book are dedicated to discussing the link between poverty and stress and to illustrating how poverty is connected to poverty-related stress. These chapters also address social, moral, and political implications and offer insights into coping, resilience, and possible responses.

In this chapter, we start by tracing how poverty has become a subject of stress research, and how the experience of poverty and the experience of stress go hand in hand.

Bringing Poverty into the Stress Research Picture

In previous chapters, we have described the processes and patterns of how organisms (single-celled as well as higher organisms such as human beings) respond when exposed to "stress," that is, to a kind of impact strong enough to severely challenge an organism's functioning or even to threaten its survival. As we have also seen, the **stress response** itself can lead to damaging as well as protective outcomes, depending on the intensity and duration of the stress experience and the resources available with which to respond. In its protective sense, a stress reaction enables an organism to adapt to the **stressor**, to manage or defeat it, allowing it to reestablish its well-balanced state of being after the impact is overcome. This balanced state can equal the level before the impact, or it can reach a different, higher level through a process of adaption—a crucial process that, using the words of Hans Selye, occurs in life "many, many times. Otherwise we could never become adapted to all the activities and demands which are man's lot" (Selye 1985). There is two-way communication between the brain and the body that both protects organisms in a situation of acute stress by activating processes in the autonomous nervous system, the **endocrine system**, and the immune system (McEwen 2005), and also enables the organism to develop and adapt. In Selye's "**General Adaptation Syndrome**" (**GAS**) model, as described in Chap. 2, this stress-driven process of development is equivalent to the alarm reaction and the stage of resistance, the first two (of three overall) stages (Selye 1985). If, however, the stressor exceeds the organism's regulatory capacity for too long, the impact cannot be overcome by these processes, and the organism remains out of balance, in an adverse state that Selye called "**distress**" and for which Bruce S. McEwen coined the term "**allostatic overload**" (Selye 1976; McEwen 2005; McEwen and McEwen 2017). Allostatic overload can lead to severe and often lasting (sometimes even lethal) effects on health, both physically and mentally. In Selye's model, this would be GAS's third stage, that of **exhaustion**.

We argue that poverty often results in a chronic state of allostatic overload. In this chapter, we will present evidence for this correlation from stress research, while in the following chapter we will present firsthand accounts exemplifying the experience of poverty-related stress.

Early stress research was able to show prevalent patterns of stress response under laboratory conditions, for instance in tests in yeast cells or in higher organisms such as rodents. Stress in those experiments was induced primarily through traumatic physical intervention. In at least one thread of subsequent research, the laboratory rooms as well as their experimental subjects (such as yeast cells and rodents) were left behind. The focus broadened to human life, including environmental and social conditions and (major and minor) disruptive life events. Such approaches, e.g., the **Social Readjustment Rating Scale** (Holmes and Rahe 1967), the Daily Hassles and Uplifts Scale (Kanner et al. 1981), the **Perceived Stress Scale** (Cohen et al. 1983), or the **Trier Inventory for the Assessment of Chronic Stress** (Schulz and Schlotz 1999)—described in depth in Chap. 6—opened stress research to stressors beyond acute physical and mental trauma, recognizing also potential entry points for stress resulting from everyday life. This touched off a large body of studies making use of lists of potential minor and major everyday stressors and assessment scales as well as discussing the significance of various stressors. Quite soon such discussions also included the question of whether social status might influence the experience of stress in any way.

In the early 1980s, in their discussion of the two then most popular modes of stress measurement ("daily hassles and uplifts" versus "major life events"), Lazarus and colleagues stated that generally, the number of undesirable life events one is likely to experience is not necessarily dependent on whether one belongs to a lower or higher "social class." However, they continued, living in disadvantaged circumstances may make it likely that a person experiences "more high-impact events of an undesirable nature" (Kanner et al. 1981).

Around the same time as the work of Kanner, Lazarus, and colleagues, and quite in line with their insights, Marc Fried also emphasized the impact of particular daily life events, such as changes, demands, and threats that result in adverse, unfavorable conditions; he pointed in particular to the importance of the experience of deprivation, especially when ongoing and even ceaseless. Fried coined the term "**endemic stress**" for such repetitive events and persistent social conditions, which he aimed to distinguish from the experiences of "acute stress" or "acute crisis" (nonetheless acknowledging that a clear distinction would be difficult if not impossible to make, as acute stress quite often could be the starting point for experiencing "chronic" or "endemic" stress; see also Turner-Cobb and Katsampouris 2019). Endemic stress, as Fried describes it, "is the phenomenon of persisting or increasing scarcity, perduring conditions of loss or deprivation, and continuing experiences of inadequate resources" (Fried 1982). Fried acknowledges that in general the origins as well as the respective nature of such events and experiences may vary considerably, yet he points out that "the most frequent and most trenchant determinants of endemic, population-wide stress in modern democratic societies are likely to be economic in origin" (Fried 1982). In this, Fried supports a strand of stress research with attention to life events and social conditions that occur in the context of low socioeconomic status and to life circumstances shaped by deprivation and poverty, which often tend to be much more enduring than episodic. In such circumstances, importance is also attached to what are sometimes called "non-events"—that is, desirable events or changes that do not happen. In 1974, Gersten and colleagues had already focused on non-events, describing them as events that may with good reason be expected and hoped for on the individual level, yet often are also common or even normative for a certain group (Gersten et al. 1974; see also Antonovsky 1990 in his critique of life-event approaches and Wheaton et al. 2013 in their overview of ways of assessing stress). We might think here of events out of a person's control, such as not getting a job applied for or not receiving recognition from a teacher despite relevant achievements. Remarkably, Fried explicitly acknowledged "poverty and powerlessness" as among the most prevalent causes for and backdrops to stressful experiences and has emphasized the need for stress research to address "conditions of chronic deprivation from which large segments of the population [suffer]" (Fried 1982). Here, poverty explicitly comes into the picture of stress research.

Poverty Causes Stress

To put it more generally, but in line with Lushchak's definition of **oxidative stress** (2014) (see Chap. 1), stress can be described as an organism's reaction to environmental demands exceeding its regulatory capacity. Chronic stress or allostatic overload is rife with situations of repeated, lasting, cumulating overstrain. Poverty means overstrain. Poverty means stress.

Poverty, as we argued in our introduction (Chap. 1), can be defined as a deprivation of tangible and intangible resources, as a situation of "lack." What is more, poverty can be defined as a situation of "cumulative lacks": lack of income, lack of living space, lack of reliable and available supportive social contacts—or in other words, "**social capital**" (especially what is called "bridging social capital," which helps a person move beyond her established social spheres)—lack of access to the labor market, health care, and the legal system. Deprivation of resources together with these cumulative lacks make high demands on a person or a family system. Not being able to pay one's bills, being in need of living space for the family but not being able to secure it, being in desperate need of medical care and not having access to it—these are the dynamics of demands exceeding the coping and responsive resources of a person; to put it another way, it is exceedingly stressful to fall behind on bills, to not be able to provide for the family, or to live with untreated medical conditions. Poverty causes stress. Living with persistent poverty damages one's health. Poverty then implies "toxic stress," a term describing a state similar to the concepts mentioned earlier of "distress" or "allostatic overload."

Aaron Antonovsky has pointed out, however, that it is not the stressor load on its own that is toxic or pathogenic, "but the inability to resolve" it (Antonovsky 1990). This is especially true for stress resulting from poverty. Here we need to remember Fried's acknowledgment of poverty and powerlessness in particular as a source of experiencing stress. Poverty will very likely involve toxic stress, with adverse conditions and demands a person (or a family or a community) may not be able to avoid, escape, or overcome. Very probably, such experiences are not simply "personal" or "private" challenges but are grounded in social and environmental contexts and relations. Poverty leaves a person more vulnerable to entry points for various forms of stress.

A person affected by poverty is generally exposed to higher social and contextual demands than the non-poor. To give a few examples: a poor person will be stopped by the police more frequently than a non-poor person; a poor person waiting somewhere will be accused of "loitering" earlier than a non-poor person; a poor person is exposed to higher levels of noise and pollution because of their living conditions. The multitude of such stressors associated with poverty, together with the lack of resources to resolve them, constitute what in stress research is referred to as "poverty-related stress" (Wadsworth et al. 2005; Wadsworth and Berger 2006; Wadsworth et al. 2008).

In the remaining sections of this chapter, we will outline some main insights of stress research on: poverty-related stress; correlations between toxic or pathogenic stress and the experience of chronic poverty; and the likely outcomes of poverty-related stress. We draw here on various studies and experiments originating mainly in neuroscience, various threads of medical research, and health studies as well as psychology. In many of these studies, poverty is addressed not in terms of experience (as we introduced it in Chap. 1), but by using the rather technical indicator "low **socioeconomic status**" (low **SES**). Low SES is usually identified via low income, low education, and/or low occupational status; such data are accessible and assessable rather easily, e.g., through simple questionnaires. Data representing "experiences," however, require different—and more complex—approaches, such as qualitative interviews or autobiographical narratives. In the following chapter, we will present firsthand accounts given by people affected by poverty from their very own experiences. We do so in order to complement the scientific studies we use on stress and low SES, i.e., living conditions framed by poverty, and thus to put flesh on the bones of the central findings these studies provide.

In a nutshell, there is strong evidence that poverty-related stress causes health problems generally and in terms of mental health problems, specifically (Wadsworth 2012). Low socioeconomic status is shown to be associated with a number of "typical" health issues,[2] "because of the stress of living with less money than one needs" (DeCarlo Santiago et al. 2011).

Research into Stress and Low Socioeconomic Status: General Notes

As we have already suggested, one major finding in stress research is that chronic, toxic stress exposure is anything but equally distributed. A review of sociological work of the past several decades on the impact of stress on health outcomes has shown that, among other indicators, such as general health status, sex, marital status, or ethnic/migrant background (some of which Fried has already explicitly identified, Fried 1982), a disadvantaged socioeconomic status accounts significantly for high levels of stress exposure as well as for accompanying adverse health outcomes (Thoits 2010). Analyses from public health studies drawing on data from longitudinal health surveys corroborate the theory that social context matters: being a member of a marginalized group or a group confronted with discrimination—which often although not always goes hand in hand with low SES—significantly contributes to disadvantages concerning health in general as well as harmful stress responses in particular (Thoits 2010; Fuller-Rowell et al. 2012; Zilioli et al. 2017). Zilioli and colleagues found that perceived discrimination plays a major role in middle-aged adults' allostatic load (Zilioli et al. 2017). Low socioeconomic status, discrimination, and stress are linked to unpromising outcomes that in turn are likely to reinforce and perpetuate disadvantage and deprivation, and thus also bring about further exposure to stress and other related health impairments.

To have available "less money than one needs" (DeCarlo Santiago et al. 2011) also converges with the experience of stress in regard to the **agency** one generally has or, rather, has not, and the choices one has or has not. This includes agency and choices regarding the quality of life as well as

---

[2] We have already discussed some examples, e.g., **major depression**, in Chap. 4, "Oxidative stress and the brain."

neighborhood conditions, often touching on what has been termed "neighborhood disadvantage" (Attar et al. 1994; DeCarlo Santiago et al. 2011; see also Sullivan et al. 2019). In fact, research has identified six major cumulative risk factors for poor families that lead to permanent high stress levels: crowding in the home, noise level, family turmoil, separation of a child from their parents, exposure to violence, and substandard accommodation (Attar et al. 1994). Poverty not only limits resources but also restricts the choices a person (or family, or community) is able to make about such conditions and surroundings. An environment of poverty creates a context with a multitude of chronic stressors (Evans 2004; Evans and English 2002; Evans and Kim 2007). Poverty is, in other words, an unfortunate combination of high environmental demands and limited resources, including limited agency and choice.

Moreover, not only does poverty limit the choices people actually have, but in turn, where choices are to be made, the pressure of stress, especially chronic stress, may have consequences for people's decision-making. In their review of research findings, Johannes Haushofer and Ernst Fehr provided evidence that suggests that chronic stress changes people's behavioral preferences: they are less willing to take risks, less willing to adopt new technologies, less able to adopt a long-term perspective, and less willing to forgo an immediate if low income in favor of higher future incomes. Such preferences are probably disadvantageous in regard to improving a person's future situation or eventually escaping poverty (Haushofer and Fehr 2014). However, as the authors emphasize, such findings should by no means be misinterpreted as "blaming the poor" for their situation. Rather, an environment of poverty (including dysfunctional institutions, exposure to violence and crime, exposure to noise and pollution, distressing living conditions, and poor access to health and educational systems, to point out just a few likely daily challenges) may have detrimental psychological effects and may shape people's minds in a disadvantageous way, in the worst case forming a vicious circle further deepening poverty (Haushofer and Fehr 2014; Rawlinson 2013; Mani et al. 2013).

Anandi Mani and colleagues have shown that even poverty-related stress induced in an experimental set-up *under laboratory conditions* can impede mental processes (Mani et al. 2013; see also Spitzer 2016). In one of their experiments, participants with low to high incomes were required to think about financial demands based on fictitious yet quite common scenarios that all included a financial challenge, such as a necessary car repair; the "costs" of each respective scenario that participants were confronted with varied from low ("easy scenario") to high ("hard scenario"). Participants were given time to decide how they would eventually solve the problem; in the meantime, they had to perform two tasks measuring their cognitive function. The experiment showed that in the easy scenario thinking about financial decisions did *not* impede cognitive function, whereas, in the hard scenario, people who had self-reported low-income status performed significantly worse than participants with higher incomes. For those with low incomes in their real lives, to be triggered to think about (even fictitious) financial problems obviously led to a mental demand too high to leave enough resources available for additional cognitive tasks. Another similar experiment Mani and colleagues conducted with Indian sugarcane farmers *in the field* confirmed their findings under non-laboratory conditions with naturally varying financial concerns and resources. Although some limitations were acknowledged (e.g., not considering possible nonmonetary circumstances affecting participants' decisions and cognitive performance or possible effects of earlier adverse experiences, such as childhood poverty), both experiments show the grave impact of immediate financial strain in adults and its consequences in terms of "bad decisions," "poor performance," and more generally, "counterproductive behavior" (Mani et al. 2013). Such and similar outcomes may not only impair a person's actual *status quo* but may also leave her vulnerable to further social discrimination, judgment, and adverse public opinion. We will discuss later how public perceptions can contribute to poverty, resulting in reinforced stress (see Chap. 10).

People are shaped by and adapt or change under the pressure of chronic, toxic stress. However, we have to bear in mind that adults do not experience the (chronic) stress response as "islands." They are part of social networks, they are part of families, and with some probability in the course of their life, they may become parents. If they suffer from circumstances that include high levels of stress exposure, such as conditions of poverty, their children will in almost all cases be affected as well. While the negative impact of poverty in terms of chronic stress and allostatic overload on health and development outcomes appears in humans of any age, at younger ages it turns out to be particularly harmful. Poverty-related stress takes effect at all ages during childhood and adolescence, yet it is particularly detrimental during infancy and even during pregnancy, doing harm to the yet unborn child. It is important to point out that poverty-related stress has the potential to set directions very early in life—and sometimes irreversibly—in regard to what a person is likely to experience and to achieve and, not least of all, to be able to *do* and *be* in the course of her life. Poverty and poverty-related stress shape lives from the very beginning.

Stress research literature in this area has emphasized at least three important topics: (1) the crucial role of *prenatal maternal stress* for the child's development before and around birth and its lasting effects during childhood and beyond; (2) the grave influence of stress exposure on *parenting practices*; and (3) the detrimental effects of experiencing poverty at a young age and of growing up in a stress-loaded environment *for the child's well-being and thriving*. The adverse effects of chronic stress and allostatic overload seen in general are reinforced when it comes to inescapable stress in relation to poverty. In the following sections, we present some core findings on the three topics highlighted above, well aware that they have to be understood as interactive.

Pregnancy, Early Childhood, Stress, and the Impact on the Child

As we have already seen, living conditions of individuals of low SES can lead to increased experiences of stress exposure, while at the same time diminishing "healthy" coping resources and strategies. Given such stress exposure during pregnancy, disadvantages in health for the child before and around birth and during early childhood can be the likely consequences (Lefmann and Combs-Orme 2014; Aizer and Currie 2014).

A keyword here is "fetal programming." For plenty of good reasons, during pregnancy, the unborn child is "programmed," in the sense of "prepared," for the conditions and challenges of life in the particular environment that he or she will be born into. The child is also programmed through her/his inherited genome and through her/his interaction with the parents. Generally, we can say that these different modes of inheritance (genetic, epigenetic, and cultural) form a closely-knit web that is nearly impossible to unravel. However, epigenetic and cultural inheritance are strongly influenced by poverty, as documented in the papers referenced below.

Thompson refers to the view that the newborn has no idea into which world and environment he/she is born; hence, the ability to quickly adapt is crucial, and the stress response is pivotal in the adaption process. The child's development, however, does not just start at birth, but during very early pregnancy via "messages" from the mother, which depend on her own capacities and resources and her own experiences. If the mother suffers from chronic stress, this experience will shape the development of the unborn child and will have a severe impact on, in Thompson's words, "how young, developing biological systems organize themselves in response to environmental signals" (Thompson 2014; Johnson et al. 2016; see also Neuenschwander and Oberlander 2017).

In Chap. 5, we have already hinted at various **intergenerational** adverse health outcomes related to chronic stress; probably the most striking example in this regard can be found in the Dutch Hunger Winter, with evidence found for in utero–acquired predisposition for **metabolic disease** (as well as for other health problems, see Thompson 2014) that can be explained by ongoing maternal malnutrition. Such adverse outcomes are not limited, though, to the field of epigenetics or to health outcomes. A number of studies from animal research as well as, in recent years, human research, show that chronic maternal stress and/or anxiety during pregnancy, such as can be caused by poverty, is also closely linked to the overall development of the child before and after birth. Lefmann and Combs-Orme explain in detail how increased levels of **HPA axis** activity (i.e., the interaction between the **hypothalamus**, **pituitary gland**, and adrenal glands) during the pregnancies of mothers under stress, along with activation of the **corticotropin-releasing hormone** (**CRH**), can affect the developing fetus. They argue that the child's development may be impaired in at least three ways. First, increased HPA axis activity has an impact on a pregnant woman's blood flow, so the unborn child may suffer insufficient provision of vital substances such as blood, water, oxygen, and nutrients. Second, high **cortisol** levels circulating from mother to child may result in premature birth and low birth weight. Third, fetal programming reacts to maternal stress hormones with consequences for the child's brain development as well as the child's own development of stress reaction mechanisms and the child's predisposition to develop type II diabetes later in life (Lefmann and Combs-Orme 2014; see also Vohr et al. 2017).

During intrauterine life, all relevant programming information about the environment is transmitted intergenerationally between mother and child. Neuenschwander and Oberlander emphasize that during this most sensitive time window, maternal prenatal stress already has an impact on the development of the **prefrontal cortex** and thus emotion regulation as well as executive function—i.e., high-order planning, reasoning, and decision-making—and cognitive processes central to self-regulatory behavior, such as inhibitory control, attentional control, and working memory (Evans and Schamberg 2009; Kim et al. 2013; Neuenschwander and Oberlander 2017; see also Johnson et al. 2016; Lefmann and Combs-Orme 2014).

There is also evidence of changes in the development or growth of the **amygdala**, prenatally and around birth. The amygdala, a part of the limbic system, is an important part of our emotional sensorium and plays a central role in receiving and processing sensory information. It is responsible for emotions such as fear and anxiety as well as positive emotional reactions, and it is also involved in the HPA axis reactivity to stress (Buss et al. 2012; Dowling

2018). Although not all is fully understood in this regard (Buss et al. 2012), there are a number of studies that find potential changes in the child's amygdala due to an enhanced cortisol level in pregnant women in chronic distress, and to such changes' detrimental consequences: maternal prenatal stress may result in children more often growing up chronically alert. In other words, children of mothers highly or chronically stressed during pregnancy are likely to stay in a constant alert mode, and so their HPA axis is activated more easily, which means that they are overly sensitive to potentially threatening situations; however, at the same time, they will probably not always be accurate in assessing situations as threatening or not. This condition can have impeding effects on children's development and their ability to thrive mentally and socially. It is also a burden for later life, since such children also appear to be more prone to anxiety disorders and depression as adults. Although all of this is not limited to the specific stress related to poverty, children growing up in deprived conditions and poor environments will much more likely be exposed to stress than their better-off peers (Evans and Kim 2013; Thompson 2014; Lefmann and Combs-Orme 2014; Johnson et al. 2016; Aust 2017). Evans and Kim term this "elevated chronic stress," and they identify it as a further pathway (joining influences during pregnancy and those from parenting and the environment) for explaining why child poverty is so harmful to the development of young children, with further consequences reaching well into adolescence and adulthood.

Other research shows that the effect of maternal or parental poverty-induced stress during pregnancy and after birth can lead to reduced development or growth of the **hippocampus** (Luby et al. 2013; Shonkoff and Garner 2014; Blair and Raver 2016). As with the amygdala, the hippocampus is part of our core emotional sensorium and is responsible for general achievements in learning and for memory. Findings on stress interfering with hippocampus development appear particularly critical, as this impact on the brain structure may affect opportunities and options in future life. As we have pointed out before, poverty-related stress and its impact on children's development may be crucial to what a person is likely to be able to *do* and to *be* in the course of her or his life. In terms of the impact of prenatal stress on decreased hippocampus growth, it is noteworthy that poor development and resulting outcomes associated with low childhood SES and poverty exposure appear to persist, even if the children overcome poverty and emerge into adulthood with a significantly higher socioeconomic status. This finding has also been corroborated by a study drawing on Scottish longitudinal data from around 250 people born in 1936, who in 1947 in the Scottish Mental Survey underwent a mental ability test (including verbal reasoning and numerical and spatial abilities) and were followed up more than 50 years later with **MR brain imaging** (Staff et al. 2012; see also Johnson et al. 2016). Results showed a significant correlation between low childhood SES and reduced hippocampal volume in (late) adulthood.

These examples give some idea of how harmful maternal stress during pregnancy is likely to turn out to be, not only for the mother's health but also, and massively, for the child's health and development. Many questions remain unanswered in this still relatively young area of research—questions that would require further long-term studies in order to better understand causalities between poverty and poverty-related stress during the course of life and especially in early life. According to Evans and Kim (2013), more research is needed on whether "early deprivation becomes embedded in the organism, in essence creating a scar that continues to fester, or whether repeated experiences of poverty over life influence human potential in a cumulative manner." As they continue: "these two perspectives are not mutually exclusive—there could be critical periods for poverty exposure as well as alterations in subsequent developmental trajectories in relation to subsequent experience of disadvantage and/or salutary phenomena (e.g., upward social mobility)" (Evans and Kim 2013).

Also, as existing longitudinal studies suggest (such as those cited on the Dutch Hunger Winter or the Scottish study cited above by Staff and colleagues), effects of experiencing poverty and poverty-related stress, especially in early life, may in some cases (at least to some degree) be reversible while others are not. In the cases above, we have seen that the early developmental stages of life are highly critical periods, and research also shows that upward mobility in later years per se does not cure all stress outcomes. These findings will be of importance when we think about possible social and political implications and ways of coping and resilience in Chap. 12.

Johnson and colleagues, in line with Lefmann and Combs-Orme, suggest the style of parenting as an important mediator (Johnson et al. 2016; Lefmann and Combs-Orme 2014). We thus proceed in this chapter with a few insights into parenting under adverse conditions and particularly under the stress of poverty. To start with, it is necessary to emphasize that parenting practices in this sense are not simply or primarily individual, personal, "good" or "bad" choices and preferences. Rather, and maybe to a considerable degree, they are dependent on social, environmental, and economic conditions. When we think about parents affected by poverty conditions, stress is likely to be among the typical important factors. Parenting practices in such a situation have to be seen as influenced by the stress reactions and adverse effects described in previous chapters.

Parenting and Family Life under Conditions of Poverty and Poverty-Related Stress

Low SES and poverty-related stress experiences not only come into play during pregnancy, but also after birth, in parenting, and during childhood, and early childhood in particular. La Placa and Corlyon (2016) write that "evidence suggests an intricate relationship between complex and mediating processes of income poverty, parental stress, disrupted parenting practices and neighborhoods and environments," also affecting diminished social networks (Evans et al. 2008). Parents from low SES backgrounds and under economic distress are repeatedly reported to more likely show what is called "disrupted" or "harsh parenting," to more likely show authoritarian as well as inconsistent parenting practices than more affluent parents. These low-SES and stressed parents also tend to be less responsive and to show less warmth and less supportive behavior toward their children. Additionally, they are reported to be less able to provide their children with cognitively stimulating environments, including age-friendly stimulating toys and formal or informal learning and reading material that might need to be purchased. It appears that they talk and read less to their kids than parents of higher SES, which results in an impoverished language environment for the child (Evans and Kim 2013; Johnson et al. 2016; La Placa and Corlyon 2016). All of this leads, additionally, to early and even prenatal entry points of risk and to a higher risk for a child not only to stay behind age-appropriate stages of cognitive and emotional development but also to develop mental disorders. Research findings also corroborate a potentially harmful impact on the adaptation of the child's own stress mechanism, often beyond the critical development periods we have referred to and reaching well into adulthood, and on the development of the immune system (Evans et al. 2008; Pascoe et al. 2016; Provençal et al. 2019; Boyce 2012). These impacts lead to further development risks due to an increased tendency to fall ill.

It is important for us to emphasize that "poor parenting of poor people" is more than a social cliché, but can be the result of physiological reactions, first on the part of the parents, and consequently also impacting their children. Zalewski and colleagues have reported altered patterns of cortisol secretion among preschool children of single parents with low socioeconomic status; they also suggest that poverty-related effects on parenting have, in turn, negative effects on the working of the HPA axis as well as the development of effortful control in children (Zalewski et al. 2012). Caregiver stress (and caregiver depression as well as other mental health problems) is more frequently experienced by children from poor households (Henninger and Luze 2014; La Placa and Corlyon 2016). Furthermore, Lareau has shown that middle-class parents are more likely to invest in "concerted cultivation" parenting (as opposed to "natural growth" parenting) than parents struggling with poverty; "concerted cultivation" provides structured activities, active interventions, and an emphasis on language development (Lareau 2011). In this sense, family stress, particularly poverty-related family stress, and its outcomes form a vicious circle, often exacerbating the situation of parents and their affected children.

Though low SES or low income enhances the risk of parenting practices detrimental to child development (and, one has to add, the possibly limited availability of feasible options around parenting practices), La Placa and Corlyon emphasize that there is no straightforward pathway between poverty per se and "parenting deficits." One reason for this is that "different people respond differently to adversity" (La Placa and Corlyon 2016); another reason is that in different cases "poverty" is not one and the same experience. There is evidence that the poverty-stress nexus strongly depends on the specific circumstances in which a person (a parent, a child, a family) actually lives. However, poverty can shape circumstances, and there is clearly more likelihood that poverty shapes circumstances for the worse. For instance, as we have already seen, low income is most often linked to having little or no choice about certain living conditions, including neighborhood and environment. Children who grow up in poverty are more likely than their better-off peers to experience psychosocial stressors, such as family turmoil, disruption, adverse social conditions, crowding, noise, and substandard housing (Evans and English 2002). The immediate environments of children from low-income families carry many risks—playing on the street can be more dangerous than playing in a fenced backyard, living in an apartment block with many parties in crowded apartments can be more fraught with risk than living in a single-family home, experimenting with "found stuff" can be more dangerous than using specifically purchased material.

Pascoe and colleagues summarize the idea that children in poverty (and their families, one may add) "face a daunting array of psychosocial and environmental inequities that undermine their healthy development" (Pascoe et al. 2016). The problem is not per se that one has to face challenges and difficulties; as many scholars note, this is inherent in life regardless of affluence or poverty or somewhere in between. The problem is the permanent exposure to multiple stressors with very little chance of escaping them, to "resolve them," as Antonovsky put it. The problem, to refer to Undine Zimmer whom we met in the introduction, is the lack of a "pause" key (Zimmer 2013).

For Pascoe and colleagues, this appears to be the "unique, critically important feature of the environment of children growing up in poverty" (Pascoe et al. 2016).

The Experience of Poverty-Related Stress and Its Impact on Childhood

Childhood poverty is harmful because it exposes children as vulnerable agents to stressful environments in very important years of their lives, probably even before birth. The naïve illusion that children who grow up in poverty may not understand, feel, or react to the poverty-related stress is just that, a naïve illusion (Wadsworth et al. 2008). "Children as young as 6 years old report significant adult-like worries about their family's financial situation, including concerns about where they will obtain their next meal, and whether they will have enough money to pay for the heating bill in the winter" (Alkon et al. 2012). Adverse childhood experiences hit ill-prepared children who cannot properly adapt, which turns an experience of stress into an experience of toxic stress. The toxic stress response results from a disruption of the circuitry between the neuroendocrine and immune systems, with effects on multiple biological systems (Bucci et al. 2016; Condon et al. 2018). As we have already seen, poverty and poverty-related stressors have various negative effects, including for example, on the HPA axis activity in young children (Mills-Koonce and Towe-Goodman 2012). (To recall: the HPA axis has been identified as a key element in behavioral processes, both cognitive and emotional, which constitute the basis for long-term negative effects.) In all likelihood, toxic stress experienced during childhood will have lifelong effects. Childhood poverty has pervasive negative effects due to its exposing children to chronic stressors. Such exposure has been associated with consequential changes in the amygdala, hippocampus, and prefrontal cortex regions crucial for cognitive development and emotion regulation, not only before birth (as discussed earlier) but also during childhood. In other words, there is a worrisome sustainability and persistence in poverty-related stress experienced during childhood (Duncan et al. 2010; Espejo et al. 2006; Huntington 2019). Children of poor parents are more likely to be poor as adults than other children because they experience early-childhood adversities, which result from social structures and relationships. These adversities impact children's bodily systems and brain development through recurrent stress (Evans and Schamberg 2009).

Chronic physiological stress leads to the "stressing out the poor" phenomena (Evans et al. 2011). Poor children are exposed to risks, and this exposure leads to stress. If this is the "normal condition," stress becomes chronic and toxic. Such a condition can result in a number of adverse effects, including, e.g., impediments to education: "toxic stress directly hinders poor children's academic performance by compromising their ability to develop the kinds of skills necessary to perform well in school" (Evans et al. 2011). Insights into the school experience of poor children have shown that the way a poor child experiences schooling is quite different than how children from better-off families experience it. As we have seen, poor children are exposed to parenting practices that are sometimes adverse to their development: they will receive lower cognitive stimulation (such as reading), they will more likely be exposed to toxic materials, and they will receive less adequate nutrition. These risks come together and generate cumulative disadvantage dynamics that account for the "achievement gap" between poor and non-poor children, i.e., the disparity in academic performance between socioeconomically challenged students and their peers (Evans et al. 2011; Huntington 2019). Individual stressors tend to aggregate in persistent conditions of poverty and social exclusion.

"Multiple stressor exposure" is a major factor in adverse childhood experience. Children growing up in poverty—not just their parents—are exposed to stressful environments with an overwhelming array of demands (social, physical, psychological) that put pressure on their adaptive capacities. Gary Evans and Kimberly English argue that it is specifically this accumulation of exposure to multiple stressors on different levels that constitutes the toxic stress associated with childhood poverty. Chronic stress exposure has deep-reaching negative effects and "may disrupt task persistence and produce disequilibrium in self-regulatory behavior" (Evans and English 2002). Challenges in socioemotional adjustment caused by stress—i.e., external factors outside of a person's ability to change—may lead to a disruptive and deviant behavior that can have further exclusionary effects that will likely include discrimination, which can again be a source of—internal—stressful experiences, even in childhood. Gary Evans has shown in a number of publications that child poverty leads to toxic stress, which—together with inadequate cognitive stimulation and parenting styles that for reasons discussed above do not encourage achievement—directly hinders poor children's academic performance. Together with many other pieces of the puzzle, this may lead to a vicious circle, thus exacerbating and consolidating poverty.

Experiencing chronic, toxic stress often involves adverse behavior in the sense of adverse ways of "coping" or "compensation," not least due to the lack of healthier alternatives, the lack of options and choices. Quite often poor children and adolescents experience their parents' unhealthy ways of dealing with stress, such as yelling, violence, smoking, drinking, and substance abuse. It is not surprising that given probable limitations to self-regulatory behavior due to stress exposure, the children may also adopt similar strategies. Additionally, as Joaquina Palomar Lever has shown in a study from Mexico City working with middle class, moderately poor, and extremely poor people, poor people tend to falsely minimize or avoid stressful situations which furthermore lowers the probability of solving their problems; they tend to use more emotionally focused coping strategies. With all this said, we should remind ourselves that patterns of coping are established for good reasons and, much more importantly, cannot be easily "unlearnt" (Lever 2008).

Concluding Remarks: Bringing Poverty More into the Picture

The undeniable fact that poverty has psychological consequences has a simple implication: poverty is not a problem that can be "fixed" like a car that needs repair; we are dealing with the frailty and complexity of human beings and the deep impact of experiences, memories, and external conditions. This perspective has to be taken into account when reflecting on coping strategies and responses. When thinking about such strategies and interventions, we also need to understand, as well as possible, what we are actually talking about when we talk about "poverty." We need to understand that "poverty," for the people concerned, is not primarily a question of statistics (including statistics on low SES) but rather a question of adverse, stressful experiences. While the impact of poverty-related stress in terms of outcomes seems well addressed and acknowledged in research, many studies dealing with stress and "poverty" (mainly understood statistically as low SES) hardly dip into the actual meaning of poverty, the experiences of and exposures to hardships or the limitations faced by the people affected. Yet they need to be seen and heard in order to provide the full picture.

In various studies of stress and poverty that we have consulted, the term "poverty" or "low SES" is not further elucidated; common indicators of low socioeconomic status are in fact used as proxies for characterizing a certain environment, yet there is hardly any or only little hint of what "poverty" or "low SES" actually means in terms of experiences. Accordingly, in a state-of-the-art review, Sara B. Johnson and colleagues point out that low SES indicators "provide little insight into how individuals actually experience poverty," as well as noting, conversely, that such examinations, guided by "technical" low-SES-indicators, ignore the fact that living marginally above a statistical poverty line does not per se prevent people from experiencing deprivation and lack of crucial resources—or in other words, poverty and poverty-related stress (Johnson et al. 2016). Given this gap in understanding what "poverty" predominantly means, not only in research on stress and poverty but also in public discourse, we aim to shed some light on experiences of poverty in subsequent chapters.

We will do so by presenting "case studies" of poverty in the sense of first-hand accounts by people who tell about enduring time spent living in poverty and the implications of having grown up in poverty, even if they have managed to overcome their state of deprivation in adulthood. We will discuss these accounts together with insights from participatory poverty research projects.

We have seen in this chapter that poverty takes effect as a chronic, possibly toxic stressor. Considering the case studies that follow, we will aim to identify some "typical elements" that make poverty a chronic, endemic, even *toxic* stressor.

As seen earlier, stress exposure can occur in a way that may be tolerable or manageable for the organism or person affected, and thus may even lead to reactions of growth. That means that stress exposure can occur in the form of stimuli that initiate an organism's or a person's healthy development—in Hans Selye's terms, we would speak of "**eustress**" here. However, and again in line with Selye's arguments, our case studies will make clear that the chronic or endemic stress exposure of poverty experiences can hardly be called "eustress." Rather, such experiences are likely to constitute severe distress or even toxic stress: toxic in the sense that it alters pathways for the worse before even a single action or decision has been made by the person herself.

References

Aizer A, Currie J (2014) The intergenerational transmission of inequality: maternal disadvantage and health at birth. Science 344(6186):856–861. https://doi.org/10.1126/science.1251872

Alkon A, Wolff B, Boyce WT (2012) Poverty, stress and autonomic reactivity. In: King R, Maholmes V (eds) The Oxford handbook of poverty and child development. Oxford UP, Oxford. https://doi.org/10.1093/oxfordhb/9780199769100.013.0012

Antonovsky A (1990) A somewhat personal odyssey in studying the stress process. Stress Med 6:71–80

Attar BK, Guerra NG, Tolan PH (1994) Neighborhood disadvantage, stressful life events, and adjustment in urban elementary-school children. J Clin Child Psychol 23:391–400

Aust S (2017) Frühe Stresserfahrungen und die Entwicklung emotionaler Fertigkeiten. Individuelle Unterschiede, neuronale Grundlagen und protektive Faktoren. In: Brisch KH (ed) Bindung und emotionale Gewalt. Klett-Cotta, Stuttgart, pp 123–143

Blair C, Raver C (2016) Poverty, stress, and brain development: new directions for prevention and intervention. Acad Pediatrics 16(3S):S30–S36

Boyce WT (2012) A biology of misfortune. Focus 29(1):1–6

Bucci M, Silvério Marques S, Oh D, Burke Harris N (2016) Toxic stress in children and adolescents. Adv Pediatr 63:403–428. https://doi.org/10.1016/j.yapd.2016.04.002

Buss C, Poggi DE, Shahbaba B, Pruessner JC, Head K, Sandman CA (2012) Maternal cortisol over the course of pregnancy and subsequent child amygdala and hippocampus volumes and affective problems. PNAS 109(20):E1312–E1319. https://doi.org/10.1073/pnas.1201295109

Cohen S, Kamarck T, Mermelstein R (1983) A global measure of perceived stress. J Health Soc Behav 24(4):385–396. https://doi.org/10.2307/2136404

Condon EM, Sadler LS, Mayes LC (2018) Toxic stress and protective factors in multi-ethnic school age children: a research protocol. Res Nurs Health 41(2):97–106. https://doi.org/10.1002/nur.21851

DeCarlo Santiago C, Wadsworth ME, Stump J (2011) Socioeconomic status, neighborhood disadvantage, and poverty-related stress: prospective effects on psychological syndromes among diverse low-income families. J Econ Psychol 32(2):218–230. https://doi.org/10.1016/j.joep.2009.10.008

Dowling JE (2018) Understanding the brain: from cells to behavior to cognition. Norton & Company, New York, London

Duncan G, Ziol-Guest K, Kalil A (2010) Early-childhood poverty and adult attainment, behavior, and health. Child Dev 81(1):306–325

Espejo EP, Hammen CL, Connolly NP, Brennan PA, Najman JM, Bor W (2006) Stress sensitization and adolescent depressive severity as a function of childhood adversity: a link to anxiety disorders. J Abnorm Child Psychol 35:287–299. https://doi.org/10.1007/s10802-006-9090-3

Evans G (2004) The environment of childhood poverty. Am Psychol 59(2):77–92. https://doi.org/10.1037/0003-066x.59.2.77

Evans G, English K (2002) The environment of poverty: multiple stressor exposure, psychophysiological stress, and socioemotional adjustment. Child Dev 73(4):1238–1248. https://doi.org/10.1111/1467-8624.00469

Evans G, Kim P (2007) Childhood poverty and health: cumulative risk exposure and stress dysregulation. Psychol Sci 18(11):953–957. https://doi.org/10.1111/j.1467-9280.2007.02008.x

Evans GW, Kim P (2013) Childhood poverty, chronic stress, self-regulation, and coping. Child Dev Perspect 7(1):43–48. https://doi.org/10.1111/cdep.12013

Evans G, Schamberg M (2009) Childhood poverty, chronic stress, and adult working memory. Proc Natl Acad Sci 106(16):6545–6549. https://doi.org/10.1073/pnas.0811910106

Evans GW, Boxhill L, Pinkava M (2008) Poverty and maternal responsiveness: the role of maternal stress and social resources. Int J Behav Dev 32(3):232–237. https://doi.org/10.1177/0165025408089272

Evans GW, Brooks-Gunn J, Kato Klebanow P (2011) Stressing out the poor: chronic physiological stress and the income-achievement gap. Community Investments 23(2):22–27

Fried M (1982) Endemic stress: the psychology of resignation and the politics of scarcity. Am J Orthopsychiatry 52(1):4–19. https://doi.org/10.1111/j.1939-0025.1982.tb02660.x

Fuller-Rowell TE, Evans GW, Ong AD (2012) Poverty and health: the mediating role of perceived discrimination. Psychol Sci 23(7):734–739. https://doi.org/10.1177/0956797612439720

Gersten JC, Langner TS, Eisenberg JG, Orzeck L (1974) Child behavior and life events: undesirable change or change per se? In: Dohrenwend BS, Dohrenwend BP (eds) Stressful life events: their nature and effects. Wiley, New York, pp 159–170

Haushofer J, Fehr E (2014) On the psychology of poverty. Science 344(6186):862–867. https://doi.org/10.1126/science.1232491

Henninger WR IV, Luze G (2014) Poverty, caregiver depression and stress as predictors of children's externalizing behaviours in a low-income sample. Child Family Social Work 19(4):467–479. https://doi.org/10.1111/cfs.12046

Holmes RH, Rahe TR (1967) The social readjustment rating scale. J Psychosom Res 11:213–218

Huntington C (2019) Early childhood development and the replication of poverty. In: Rosser E (ed) Holes in the safety Net: federalism and poverty. Cambridge University Press, Cambridge, pp 130–150. https://doi.org/10.1017/9781108631662.007

Johnson SB, Riis JL, Noble KG (2016) State of the art review: poverty and the developing brain. Pediatrics 137(4):e20153075. https://doi.org/10.1542/peds.2015-3075

Kanner AD, Coyne JC, Schaefer C, Lazarus RS (1981) Comparison of two modes of stress measurement: daily hassles and uplifts versus major life events. J Behav Med 4(1):1–39. https://doi.org/10.1007/bf00844845

Kim P, Evans GW, Angstadt M, Ho SS, Sripada CS, Swain JE, Liberzon I, Phan KL (2013) Effects of childhood poverty and chronic stress on emotion regulatory brain function in adulthood. PNAS 110(46):18442–18447. https://doi.org/10.1073/pnas.1308240110

La Placa V, Corlyon J (2016) Unpacking the relationship between parenting and poverty: theory, evidence and policy. Soc Policy Soc 15(1):11–28. https://doi.org/10.1017/s1474746415000111

Lareau A (2011) Unequal childhoods: race, class, and family life. University of California Press, Berkeley, CA

Lefmann T, Combs-Orme T (2014) Prenatal stress, poverty, and child outcomes. Child Adolesc Soc Work J 31(6):577–590. https://doi.org/10.1007/s10560-014-0340-x

Lever JP (2008) Poverty, stressful life events, and coping strategies. Span J Psychol 11(1):228–249. https://doi.org/10.1017/s1138741600004273

Luby J, Belden A, Botteron K, Marrus N, Harms MP, Babb C, Nishino T, Barch D (2013) The effects of poverty on childhood brain development. The mediating effect of caregiving and stressful life events. JAMA Pediatrics 167(129):1135–1142. https://doi.org/10.1001/jamapediatrics.2013.3139

Lushchak VI (2014) Free radicals, reactive oxygen species, oxidative stress and its classification. Chem Biol Interact 224:164–175. https://doi.org/10.1016/j.cbi.2014.10.016

Mani A, Mullainathan S, Shafir E, Zhao J (2013) Poverty impedes cognitive function. Science 341(6149):976–980. https://doi.org/10.1126/science.1238041

McEwen BS (2005) Stressed or stressed out: what is the difference? J Psychiatry Neurosci 30(5):315–318

McEwen CA, McEwen BS (2017) Social structure, adversity, toxic stress, and intergenerational poverty: an early childhood model. Annu Rev Sociol 43:445–472. https://doi.org/10.1146/annurev-soc-060116-053252

Mills-Koonce WR, Towe-Goodman N (2012) Poverty and HPA functioning in Young children. In: King R, Maholmes V (eds) The Oxford Handbook of poverty and child Development. University of Oxford, Oxford. https://doi.org/10.1093/oxfordhb/9780199769100.013.0022

Neuenschwander R, Oberlander TF (2017) Developmental origins of self-regulation: prenatal maternal stress and psychobiological development during childhood. In: Deater-Deckard K, Panneton R (eds) Parental stress and early child development. adaptive and maladaptive outcomes. Springer Nature, Cham, pp 127–156. https://doi.org/10.1007/978-3-319-55376-4_6

Pascoe JM, Wood DL, Duffee JH, Kuo A (2016) Mediators and adverse effects of child poverty in the United States. Pediatrics 137(4):e20160340, e2–e3. https://doi.org/10.1542/peds.2016-0340

Provençal N, Arloth J, Cattaneo A, Anacker C, Cattane N, Wiechmann T, Röh S, Ködel M, Klengel T, Czamara D, Müller NS, Lahti J, PREDO team, Räikönnen K, Pariante CM, Binder EB (2019) Glucocorticoid exposure during hippocampal neurogenesis primes future stress response. PNAS 117(38):23280–23285. https://doi.org/10.1073/pnas.1820842116

Rawlinson RM (2013) A mind shaped by poverty. Veiled Threads. Regenia M. Rawlinson

Schulz P, Schlotz W (1999) Trierer Inventar zur Erfassung von chronischem Streß (TICS): Skalenkonstruktion, teststatistische Überprüfung und Validierung der Skala Arbeitsüberlastung. Diagnostica 45(1):8–19. https://doi.org/10.1026//0012-1924.45.1.8

Selye H (1976) Forty years of stress research: principal remaining problems and misconceptions. CMA J 115:53–56

Selye H (1985) The nature of stress. Basal Facts 7(1):3–11

Shonkoff JP, Garner AS (2014) The lifelong effects of early childhood adversity and toxic stress. Pediatrics 129(1):e232–e246. https://doi.org/10.1542/peds.2011-2663

Spitzer M (2016) Armut macht dumm. Nervenheilkunde 4:252–261

Staff RT, Murray AD, Ahearn TS, Mustafa N, Fox HC, Whalley LJ (2012) Childhood socioeconomic status and adult brain size: childhood socioeconomic status influences adult hippocampal size. Ann Neurol 71(5):653–660. https://doi.org/10.1002/ana.22631

Sullivan S, Kelli HM, Hammadah M, Topel M, Wilmot K, Ramadan R, Pearce BD, Shah A, Lima BB, Kim JH, Hardy S, Levantsevych O, Obideen M, Kaseer B, Ward L, Kutner M, Hankus A, Ko Y-A, Kramer MR, Lewis TT, Brammer JD, Quyyumi A, Vaccarino V (2019) Neighborhood poverty and hemodynamic, neuroendocrine, and immune response to acute stress among patients with coronary artery disease. Psychoneuroendocrinology 100:145–155. https://doi.org/10.1016/j.psyneuen.2018.09.040

Thoits PA (2010) Stress and health: major findings and policy implications. J Health Soc Behav 51(S):S41–S53. https://doi.org/10.1177/0022146510383499

Thompson RA (2014) Stress and child development. The Future Child 24(1: helping parents, helping children: two- generation mechanisms):41–59. https://doi.org/10.1353/foc.2014.0004

Turner-Cobb J, Katsampouris E (2019) Stress. In: Llewellyn C, Ayers S, McManus C, Newman S, Petrie K, Revenson T et al (eds) Cambridge handbook of psychology, health and medicine (Cambridge handbooks in psychology). Cambridge University Press, Cambridge, pp 149–153

Vohr BR, Poggi Davis E, Wanke CA, Krebs NF (2017) Neurodevelopment: the impact of nutrition and inflammation during preconception and pregnancy in low-resource settings. Pediatrics 139(s1):S38–S49. https://doi.org/10.1542/peds.2016-2828F

Wadsworth ME (2012) Working with low-income families: lessons learned from basic and applied research on coping with poverty-related stress. J Contemp Psychother 42:17–25. https://doi.org/10.1007/s10879-011-9192-2

Wadsworth ME, Berger LE (2006) Adolescents coping with poverty-related family stress: predictors of coping and psychological symptoms. J Youth Adolesc 35:57–70. https://doi.org/10.1007/s10964-005-9022-5

Wadsworth ME, Raviv T, Compas BE, Connor-Smith JK (2005) Parent and adolescent responses to poverty- related stress: tests of mediated and moderated coping models. J Child Fam Stud 14(2):283–298. https://doi.org/10.1007/s10826-005-5056-2

Wadsworth ME, Raviv T, Reinhard C, Wolff B, Santiago CD, Einhorn L (2008) An indirect effects model of the association between poverty and child functioning: the role of children's poverty-related stress. J Loss Trauma 13:156–185. https://doi.org/10.1080/15325020701742185

Wheaton B, Young M, Montazer S, Stuart-Lahman K (2013) Chapter 15: social stress in the twenty-first century. In: Aneshensel CS, Phelan JC, Bierman A (eds) Handbook of the sociology of mental health, 2nd edn. Springer, Amsterdam, pp 299–323. https://doi.org/10.1007/978-94-007-4276-5_15

Zalewski M, Lengua LJ, Fisher PA, Trancik A, Bush NR, Meltzoff AN (2012) Poverty and single parenting: relations with preschoolers' cortisol and effortful control. Infant Child Dev 21(5):537–554. https://doi.org/10.1002/icd.1759

Zilioli S, Imami L, Ong AD, Lumley MA, Gruenewald T (2017) Discrimination and anger control as pathways linking socioeconomic disadvantage to allostatic load in midlife. J Psychosom Res 103:83–90. https://doi.org/10.1016/j.jpsychores.2017.10.002

Zimmer U (2013) Nicht von schlechten Eltern. Meine Hartz-IV-Familie. S. Fischer, Frankfurt a. M.

9

The Stressful Experience of Poverty

Poverty is a complex phenomenon and can be understood in many different ways, from the varied perspectives of material and immaterial "resources," of "access" to institutions and markets, and of "freedoms" and "capabilities," to mention but a few. While poverty framed as "low **socioeconomic status**[1] (low **SES**)" draws mainly on demographic *data* (such as low level of education, employment status, and/or income), we have in the introduction of this book emphasized an understanding of poverty, at the same time more extended and in-depth, as an *experience*—an experience that both shapes people and threatens identity and **integrity.**[2]

We have previously introduced the idea that, in a number of ways, stress can be toxic. In the preceding chapters, we furthermore suggested that poverty can be a source of **toxic stress**, and also presented insights from stress research into correlations between "poverty" (or low SES) and stress outcomes in terms of development and health. Findings from SES-sensitive stress research strongly support the concept of poverty-related toxic stress. In this chapter, we want to take a "deep dive" into the experience of poverty. If we engage with people who know what it is like to live in conditions of adversity and destitution, what can they tell us about the stressful experience of poverty?

---

[1] Glossary terms are bolded at first mention in each chapter.
[2] We will say more about identity and integrity in Chap. 11; see also Bray et al. 2019, who name "identity" as one of the modifying, i.e., intensifying or mitigating, factors in the experience of poverty.

© Springer Nature Switzerland AG 2021
M. Breitenbach et al., *Stress and Poverty*, https://doi.org/10.1007/978-3-030-77738-8_9

"Speaking the 'Truth' of Poverty"

There are actually plenty of accounts available from people who exemplify what Irene Edge and colleagues call "living poverty" (Edge et al. 2004), including autobiographical statements and testimonies facilitated by participatory poverty research. Even if only referring here to *written* accounts, we are able throughout this book to draw not only on findings from research projects and experiments but also on firsthand reports from "real life." Such accounts are of immense value, and their insights offer a very special type of poverty knowledge, knowledge from within. For poverty research, it is crucial to not only "lend a voice to suffering" but to acknowledge the "voices *of* suffering in speaking the 'truth' of poverty," as Ruth Lister puts it (Lister 2021). As she emphasizes, "in addition to traditional forms of expertise associated with those who theorize and research poverty, there is a different form of expertise borne of experience" (Lister 2021). In this chapter, we center on two such substantial sources of expertise "speaking the 'truth' of poverty," complemented by insights from exemplary participatory research projects on the lived experience of poverty.

Our two central autobiographical sources stem from two people experiencing poverty in different times and different places and at different stages of their lives. In Carolina Maria de Jesus we meet a mother of three children dwelling in a favela in São Paolo in the 1950s. De Jesus wrote down her experiences and thoughts in a diary, right in the midst of her struggles. By contrast, Darren McGarvey grew up in a dysfunctional family living in an impoverished suburb of Glasgow in the late twentieth century and provides reflections while looking back on his childhood and youth and on the conditions that surrounded him. In many aspects their stories are different, yet they show a notable number of overlaps and congruence in terms of experiences. They talk about living in surroundings permeated by ongoing violence. They reflect on being deprived of a carefree childhood, and they talk about what it means to lack any space for rest and relaxation. They share incidents of humiliation, of shame and of being shamed, and the regular experience of being treated as different, of being "othered" by the non-poor, to "count for nothing" (Lister 2015; Lister 2021). Both of them, de Jesus and McGarvey, give a vivid and striking testimony of how stressful it is to experience poverty.

"I'm One of the Discarded": The Story of Carolina Maria de Jesus

Turning to the first account, Carolina Maria de Jesus's diary, *Child of the Dark*, we will be discussing a classic text, frequently used as an introduction to poverty studies. This private journal of a woman fighting poverty in 1950s São Paolo was discovered by a journalist, and an edited version was first published in 1960. The book serves as "a kind of tour guide through the usually hidden world of poverty," says Robert Levine. "Her descriptions of favela life do not romanticize its poverty, but hack away at it with a machete" (Levine 1992). It is both a testimony of poverty-induced stress and a testimony to stress **resilience**. As one reviewer commented: "The diary only covers three of Carolina's nearly fourteen years in the favela. A truly great book, it brings vividly alive a woman of astounding courage, strength, gentleness and wisdom, in her dramatic conflict with Hunger which threatens to destroy her and her children. It describes social conditions which are rightly termed the cancer of modern civilization" (Nicholas 1963).

Carolina Maria de Jesus was born in 1913 in Sacramento, Brazil. She was a Black woman with only a couple of years of formal education. She started work at the age of 16, and after becoming pregnant with a son, her life became stressful. In his preface, David St. Clair, translator of the English edition of her diary, describes her situation: "With a baby she couldn't work. He had to be looked after constantly. She heard that junkyards paid for scrap paper and so, strapping her tiny son to her back, she walked the streets of rich São Paulo looking for trash" (de Jesus 2003[3]). She gave birth to two more children by different men and struggled as a single mother in São Paulo. The first entry in the published version of the diary talks about moral stress, the stress induced by the incapability to live according to well-justified moral standards: "*July 15, 1955*. The birthday of my daughter Vera Eunice. I wanted to buy a pair of shoes for her, but the price of food keeps us from realizing our desires. Actually we are slaves to the cost of living. I found a pair of shoes in the garbage, washed them, and patched them for her to wear" (CMJ 3). "We are slaves to the cost of living," conveys a lack of freedom, dependence, constraint, and pressure. The poverty-related stress overshadows the experience of beauty: "There are so many beautiful things in the world that are impossible to describe. Only one thing saddens us: the prices when we go shopping. They overshadow all the beauty that exists" (CMJ 36). Living under such

---

[3] Going forward, we will use the abbreviation "CMJ" to refer to *Child of the Dark*, with page numbers from the paperback edition of 2003.

conditions leads into despair: "I heard women complaining with tears in their eyes that they couldn't bear the rising cost of living any more" (CMJ 86). Many aspects of life turn into a struggle. On July 20, 1958, she writes: "I felt I was as a battlefield where no one was going to get out alive" (CMJ 58). And about a year later (June 12, 1959): "My battle of the day was to fix lunch" (CMJ 166). Life in the favela is a battle, and not just for Carolina and her family: "Here in the favela almost everyone has a difficult fight to live" (CMJ 28). She finds colorful expressions to describe life as a struggle: "Hard is the bread that we eat. Hard is the bed on which we sleep. Hard is the life of the favelado" (CMJ 33).

Life in poverty is stressful: there is no peace of mind, no rest, no escape. She does not sleep well in the favela (CMJ 83), because of the noise, the suffocating air, anxiety, and hunger. "He who is hungry doesn't sleep" (CMJ 122). Hunger is "real slavery" (CMJ 23). There are fleas and bedbugs (CMJ 108, 125). Sometimes it is raining through the roof, and she has to get up in the middle of the night to alleviate the damage (CMJ 117). Sometimes there is no water or water has become very scarce (CMJ 106), sometimes—in the rainy season—there are floods and torrents (CMJ 138–139). She does not have a quiet space (CMJ 93). She cannot find tranquility. "The poor don't rest nor are they permitted the pleasure of relaxation" (CMJ 4). Poverty deprives her of inner peace: "I am not happy with my spiritual state. I don't like my restless mind" (CMJ 141). She experiences poverty as a life under constant pressure: "I've always heard it said that the rich man doesn't have peace of mind. But the poor doesn't have it either, because he has to fight to get money to eat with" (CMJ 150). The presence of the **stressors** is constant, chronic, persistent, pervasive. She feels the dynamics that Zygmunt Bauman described in terms of "wasted lives" (Bauman 2004); she writes, "I classify São Paulo this way: The Governor's Palace is the living room. The mayor's office is the dining room and the city is the garden. And the favela is the backyard where they throw the garbage" (CMJ 24). In an entry not long afterward, she records: "When I am in the city I have the impression that I am in a living room with crystal chandeliers, rugs of velvet, and satin cushions. And when I'm in the favela I have the impression that I'm a useless object, destined to be forever in a garbage dump" (CMJ 29). Again and again, she uses the term "garbage dump" to talk about the favela. Living in poverty and in the favela means living in a detested environment, in conditions she hates: "What disgusts me is that I must live in a favela" (CMJ 14).

Poverty-related stress and the constant struggle for survival change people: "The favelados are the few who are convinced that in order to live, they must imitate the vultures" (CMJ 33). This leads to anxiety-based isolation: "Who lives in the favela must try and isolate themselves—live alone" (CMJ 41). Self-isolation, in turn, aggravates the stressful situation due to a lack of social support systems. There is a lack of empathy, a lack of compassion: "Why is it that the poor don't have pity on the other poor?" (CMJ 73). Even in the case of illness, de Jesus observes, they do not help each other (CMJ 84). In addition to the lack of empathy, compassion, and mutual support, everyday life in the favela is filled with violence and fights: "I'm so used to seeing fights" (CMJ 94). Violence is the standard way of dealing with conflicts, "things that could be solved with words they transform into fists" (CMJ 44). De Jesus admits that she herself also beats her children. She has to lock them up for hours in their shack while she is looking for garbage to sell and for food. She also describes a phenomenon recounted in Oscar Lewis's studies on "the culture of poverty" (Lewis 1959), namely the loss of childhood. Children are exposed to experiences they should be protected from: "The evening in a favela is bitter. All the children know what the men are doing … with the women" (CMJ 126).

De Jesus compares the favela to hell (CMJ 153): "A favela is a strange city and the mayor here is the Devil. The drunks who are hidden during the day come out at night to bother you" (CMJ 83). The favela shapes a person, leads to a hardening of manners and heart: "Sometimes families move into the favela with children. In the beginning they are educated, friendly. Days later they use foul language, are mean and quarrelsome. They are diamonds turned to lead. They are transformed from objects that were in the living room to objects banished to the garbage dump" (CMJ 30). The favela is hell without peace, without rest, without trust, without a sense of beauty and community: "I'm … a favelado. I'm one of the discarded. I'm in the garbage dump and those in the garbage dump either burn themselves or throw themselves into ruin" (CMJ 29). She is used to racism, she is used to rejection: "We of the favela are feared" (CMJ 75), she writes. In another account, we read: "I went past the canning factory and found a few tomatoes. The manager when he saw me began to swear at me. But the poor must pretend that they can't hear" (CMJ 64). "The neighbors in the brick houses look at the favelados with disgust. I see their looks of hate because they don't want the favela here. They say the favela debases the neighbourhood and that they despise poverty" (CMJ 49).

Next to physical and social stressors, there is also the psychological burden of shame and humiliation. Carolina Maria de Jesus describes the experience of not being able "to walk about without shame" (as poverty has been described by Amartya Sen, see Zavaleta Reyles 2007); she is ashamed of her appearance, of her clothes, of her smell, of working as a garbage collector. She has no money for soap (CMJ 87, 91), and water, much of the time, is short: "If I walk around dirty it's because I'm trapped in the life of the favelado" (CMJ

35). The toxic stress she is exposed to undermines her will to live: "How horrible it is to get up in the morning and not have anything to eat. I even thought of suicide" (CMJ 92); "I'm starting to lose my interest in life," she writes. "It's beginning to revolt me and my revulsion is just" (CMJ 27). The constant stress levels make her tired, exhausted: "I'm tired of working so hard" (CMJ 169).

The most stressful factor is "food stress," the lack of food security: "I was furious with life and with a desire to cry because I didn't have money to buy bread" (CMJ 173). Hunger is her constant companion. She does not know in the morning, when she sets out to look for discarded items that she could sell, whether there will be one meal or two meals on that day, or no meal at all. This is a recurring experience: "my problem is always food" (CMJ 43). On May 27, 1958, she had no food in the morning: "I didn't have any breakfast and walked around half dizzy" (CMJ 37). On August 7 in the same year, she writes: "I got out of bed at 4 a.m. I didn't sleep because I went to bed hungry. And he who lies down with hunger doesn't sleep" (CMJ 98). And 2 days later: "I got out of bed furious. With a desire to break and destroy everything. Because I only have beans and salt. And tomorrow is Sunday" (CMJ 99). As a mother she is in a constant state of worry about her children: "A mother is always worried that her children are hungry" (CMJ 108).

In its richness and thickness, the language of the journal points to a twofold dynamic: poverty causes stress, but stress also consolidates poverty, because of health problems, violent behavior, lack of concentration, and lack of access to support systems. Poverty causes stress, and stress causes ill-health.

In a qualitative study working with 56 people living in the Highbridge neighborhood of the Bronx in New York City, the participants described a direct causal link between stress and poor health (Kaplan et al. 2013). They perceived their stress levels as a threat and an impediment to an acceptable quality of life. It is noteworthy that several sources of stress highlighted by the participants (living in a disadvantaged neighborhood, but not in a favela) are concordant with the experiences recorded by de Jesus, namely: poverty, poor housing, food insecurity, violence, discrimination and stigma, poor working conditions, and unemployment. One participant also pointed explicitly to the overwhelming quantity of issues that makes stress "toxic": "You know, poor folks always got something ain't right ... Not one thing that come, it be an avalanche" (Kaplan et al. 2013). Similarly, a large and influential participatory World Bank study has shown that the stress experienced by poor people has dire effects for households: "Households are crumbling under the stresses of poverty. The household as a social institution is crumbling under the weight of poverty. While many households are able to remain intact, many others disintegrate as men, unable to adapt to their 'failure' to earn adequate incomes under harsh economic circumstances, have difficulty accepting that women are becoming the main breadwinners and that this necessitates a redistribution of power within the household. The result is often alcoholism and domestic violence on the part of men, and a breakdown of the family structure" (Narayan 2000).

We find the same observations in the diary from São Paolo and are very likely to find similar observations in other places, too, all over the world. It is not surprising that de Jesus's diary can be utilized in political fields of action and read from a collective perspective (Piña Cabrera 2010). Accordingly, in his preface to de Jesus's book, David St. Clair writes: "Carolina is not really the main personage in her diary. It is a bigger character—Hunger. From the first to the last page he appears with an unnerving consistency. The other characters are consequences of this Hunger: alcoholism, prostitution, violence, and murder. The human beings who walk through these pages are real, with their real names, but with slight variations they could be other [people] who live with Hunger in New York, Buenos Aires, Rome, Calcutta, and elsewhere" (CMJ xiv).

From Brazil in the 1950s we turn to Scotland around the turn of the millennium. From a mother's view, we switch to the account of a boy, a young man, a son. From life in the favela, repeatedly referred to by de Jesus as "hell," we move to a deprived housing estate, which, however dirty and violent, the author still portrays as "home." First published in 2017 as *Poverty Safari* (McGarvey 2018[4]), Darren McGarvey's experiences, one might assume, would be quite different from those of Carolina Maria de Jesus. However, it becomes apparent that although their respective *situations* are not the same, the two share a number of similar, and similarly stressful, *experiences* resulting from severely deprived environments and living conditions. Both face demands that challenge or even exceed their capacities and resources. Both endure the ongoing presence of violence; the harm done to childhood; the lack of room for rest and relaxation; the experience of shame and humiliation; and the regular experience of being treated as different, of being dismissed— in other words, of being "othered"—especially when interacting with the affluent rest of society, with the non-poor societal majority.

---

[4] To refer to *Poverty Safari*, we will hereafter use the abbreviation "DMG," with page numbers from the 2018 paperback edition.

"Having No Margin for Error": The Story of Darren McGarvey

Darren McGarvey was born in 1984, to very young parents—"kids themselves" then, as the author states (DMG 199)—who both came from poor, alcoholic homes. His mother, from a severely deprived and abusive background, left the family when he was 10 and died 7 years later. The boy grew up with his father and four siblings in Pollok, a then low-income, rundown, and rather dangerous neighborhood in the southwest of Glasgow. He was, as he puts it, "invited to leave the family home" (DMG 64) shortly before his mother's death and became homeless shortly afterward (just as his mother had been not long before her death). He found himself staying with friends, other family members, or with his then-girlfriend, in a relationship that, as he says, was "showing signs of dysfunction," (DMG 66), before he was finally given the opportunity to move into a housing project for young vulnerable adults at risk of homelessness or actually homeless, as McGarvey was (DMG 66). This episode, however promising at first glance, unluckily resulted in a decade of substance abuse he has since managed to overcome.

In a disadvantaged neighborhood, the burden of too many demands weighs people down; this is true even at a young age, as it was in Pollok. The burdens people face include the experience of violence, or as McGarvey recalls, the threat of violence: "Acts of violence are terrifying, but a sustained threat of violence is sometimes much worse" (DMG 11). With similarities to findings from stress research, he states: "You adapt to the threat by becoming hypervigilant" (DMG 11). Violence is present out on the streets of Pollok, but also at home. "In a home where violence, or the threat of violence, is regular, you learn how to negotiate it from a young age" by developing various "survival strategies," in order to assess situations, but also to "manipulate" the possible abuser and "keep an abuser's anger at bay" (DMG 11). However, "these strategies only work for so long before they inevitably fail" (DMG 12), and the dreaded violent incident materializes. In a dramatic flashback, McGarvey recalls one such incident within the family home: he describes how his mother, very likely under the influence of alcohol and other substances, chased the then 5-year-old boy through the house for not staying in bed and interrupting her party with friends, probably also embarrassing her in front of her guests. "Seconds before, she had appeared to be having so much fun that it had felt safe to wind her up in front of people. Now I was trapped in my room, pinned against the wall, with a knife to my throat" (DMG 13). McGarvey concedes that it is "hard to quantify what an experience like that does to a person," but such events, "seeming strangely normal at the time" (DMG 13), certainly shape the way a person is going to perceive the world and will definitely not appease hypervigilance. "For if you are not safe in your own home … then where else could you possibly drop your guard?" (DMG 13). It may have been encounters like this one that made McGarvey attentive to people experiencing stress, and especially poverty-related stress.

Years later, working at a local library with play groups that were mainly composed of children "from the poor section of the community," a girl withdraws from the activity and turns to Darren, "visibly stressed by the sound of the other children having fun" (DMG 84). In the ensuing conversation, the girl reveals that her father is in prison because of anger and violence, that he is "scary," and that her mother is trying to get both of them away from him. In her obvious state of hypervigilance, she "has already adapted to the reality that her life is frightening" (DMG 85), and her experience is "already having a profound impact on her ability to socialise and connect, creating an isolation urge" (DMG 86). Despite being in safe surroundings, the girl is interpreting the laughter and noise of children enjoying themselves as a threat. Again, we must ask, as McGarvey does: if this girl "interprets children playing in a safe environment as a threat, then you must wonder, under what conditions could she possibly relax and feel safe?" (DMG 86–87).

From the examples of the favela diary and the participatory research project in Highbridge, we have learned about significant elements that account for poverty being an experience loaded with toxic stress: the various problems and demands imposed are overwhelming and often inescapable, and at the same time there is hardly any room for rest and relaxation. If there is one very likely constant for people living in poverty and deprivation, it is lasting pressure, chronic stress. At least in part, this overwhelming stress is very likely of a different kind for people experiencing poverty than for the "non-poor." One participant of the Highbridge study makes this very clear when she speaks to her focus group leader: "Are we talking about *your* kind of stress or *my* kind of stress? … Our world, our stress, is totally different from yours" (Kaplan et al. 2013). McGarvey's invitation to come with him on a "poverty safari" is no less than an invitation to better understand this different world, as was the "guide" offered by de Jesus "through the usually hidden world of poverty" (Robert Levine, in his afterword to *Child of the Dark*, CMJ 184).

The permanent stress of living in conditions of poverty does not remain without consequences, as McGarvey points out repeatedly in his book. It is problematic if "hypervigilance becomes your default emotional setting" (DMG 11). Then, stress "begins to alter your physiological state" (DMG 62). Stress is ongoing, and life events are not always foreseeable. The "pause" key that Undine Zimmer wished to be able to press (Zimmer 2013, see also Chap. 1), is just not there. "A vulnerable family living in constant economic uncertainty, job insecurity or subject to an inhumane sanctions regime often lacks the capacity to absorb, process and practically address life's unpredictable adversities" (DMG 96).

We have in previous chapters discussed how chronic economic, environmental, and social stress—and hence chronically increased **allostatic load**—affects the opportunities, choices, and options a person has in terms of who to be or become. This effect includes a spatial dimension: there is hardly any place "in a deprived community that is quiet enough to hear yourself think" (DMG 136). McGarvey gives a vivid account of what this means by describing the "sounds of a stressful community" that form the noisy background of a stressful everyday life: "paper-thin walls that mean you can hear your neighbours flushing toilets, boiling kettles, having sex, arguing, doing DIY, cutting their grass, revving their cars—at every hour of the day. This is not to mention the less-than-serene sound of a stressful community, and all the challenging, often frightening, behaviour it fuels; couples engaged in aggressive disputes, drunken young people shouting in the streets, strangers coming and going all day and night, not to mention the regular sound of police cars, ambulances and fire engines" (DMG 137), With his last sentence, McGarvey suggests the acts of violence always lurking and all too often emerging around the neighborhood. In his depiction, he makes obvious both what has been described in Chap. 8 as "neighborhood disadvantage" (Attar et al. 1994; Sullivan et al. 2019) as well as what has been said about the adverse effects of noise (Münzel et al. 2014, see Chap. 4) in terms of the lack of peace and quiet and the tranquility needed for concentration.

People who do not share the experience of living in poverty very likely—and understandably, as McGarvey concedes—lack insight into the ongoing stress poor people experience; those "who don't live in poverty might be detached from the struggles of those who do" (DMG 92). Detachment, however, does not imply any "neutral" view the affluent have of the poor—and vice versa, as McGarvey emphasizes repeatedly in his book. We have already seen with de Jesus how poverty leads to rejection by the better off and, for the poor, is painful in the sense of being seen through the eyes of others, misinformed or misguided as they may be. As a result, at least two things experienced by people from deprived communities in regard to society as a whole have to be addressed: first, feelings of shame and humiliation, and second, the sense of detachment—of being left behind or disconnected; both of these feelings lead to anger and resentment (Lister 2021).

Humiliation can be a consequence of the ignorance of authorities, who can worsen poor people's situations significantly by increasing the pressure they are under. In McGarvey's words: "Take the UK welfare system at present, where it appears that humiliation is being used to incentivize people into finding a job. Such an approach could only be dreamt up by people who have no idea of what being born poor is really like. What it does to your mind, body and spirit. Poverty is not only about lack of employment, but about having no margin for error while living in constant stress and unpredictability" (DMG 96). "Having no margin for error" not only applies to interactions with authorities but also within the community, the neighborhood. This is where feelings of shame come into play. McGarvey describes this with regard to his own dysfunctional family: "When you live in a troubled home, life spills out onto the street" (DMG 44). Here, distinctions one might have assumed between the experiences of having to dwell in a crowded Brazilian favela or in a home in a disadvantaged Glaswegian neighborhood ultimately fade. You cannot do anything about it: "Concerned neighbours hear your troubles through the wall" (DMG 44). Judgment looms not only from the authorities but on your very own doorstep: "You adjust to the fact that people in the community know your business and are probably judging you. Privacy becomes another elusive luxury for fancy people" (DMG 44). The problem is that the "fancy people" in this case are not always distant people of the middle or higher classes but can be your immediate neighbors or the school friends you decide to bring home. McGarvey recalls a sunny afternoon not long before his by-then severely drug-addicted mother would leave the family: "I arrived with a couple of friends in tow, to find many of the contents of our home laid out in the front garden, incinerated. I can't recall what explanation I offered my friends, though I suspect none was required. They already had some insight into how we lived." Shame and humiliation emerge where it is no longer possible for people to hide whatever they would rather keep private from others, be they friends or not. Shame and humiliation cause additional strain and can contribute to poverty-related stress. In a (poverty-induced) dysfunctional private life, it seems plausible, though, that "you become … closed-off to the dysfunction, perhaps to spare yourself feelings of shame or embarrassment" (DMG 44). The involuntary disclosure of family turmoil, especially when accompanying conditions of poverty, may then be a considerable source of shame. In McGarvey's words: "Pretending you're not poor is one thing … but concealing family dysfunction is much trickier. For one, the dysfunction might be out of your control; a parent or a sibling, for example. Second, the dysfunction may be imperceptible to you and therefore hard to hide. Dysfunction, like poverty, can lead to disfigurements which are visible to everyone but you" (DMG 44)—at least until you are forced to see your front garden through the eyes of your friends; or when you see your whole living situation through the lens of authorities, including those in charge of social benefits, whose decisions certainly have a tremendous impact on the stress of people living in poverty. Feelings of shame set in when hiding strategies fail, and a person becomes aware of how she is seen from "outside," by others.

To be labeled as "poor" means to be identified and treated as different (Lister 2015, 2021) and also to be blamed for being poor. Many people familiar with the experience of poverty describe a feeling of "being othered" by the better-off, maybe affluent fellow citizens, but paradoxically also by people living in poverty who deny being "poor" themselves (Shildrick and MacDonald 2013). We saw similar dynamics before when de Jesus complained of the lack of compassion and support among the *favelados*. There are accounts of how "being othered" feels. For instance, a participant in the study "The Roles We Play," initiated by ATD Fourth World in the United Kingdom, criticizes the derogatory ways the public, politicians, and the media speak about "the poor" as "'lazy', 'scroungers', 'feckless parents' and 'underclass'. The stereotyping of all poor people dehumanizes them in the eyes of others" (ATD Fourth World 2014). Another woman taking part in this study expresses the view that as a home caregiver depending on welfare benefits, she feels "attacked" by the public and the media. She explains how her situation is far from comfortable: "I hate the stigma, shame, insecurity and instability that come with being on benefits. I shouldn't have to feel ashamed" (ATD Fourth World 2014). Another international participatory poverty research study facilitated by ATD Fourth World, focusing on the "Hidden Dimensions of Poverty," finds that "'othering' (saying or thinking 'We're not like those people') is commonplace" (Bray et al. 2019). Experiencing stigma and othering is also a topic in stress research, and as we have already seen, perceived discrimination can not only lead to a number of severe physical and mental disorders but can also compromise a person's behavioral control and their ability to regulate emotions, e.g., control negative emotions such as anger (Thoits 2010; Zilioli et al. 2017; Fuller-Rowell et al. 2012; see also Chap. 8.). Zilioli and colleagues give the example—very well fitting McGarvey's accounts—of displaying deviant behavior, such as substance abuse, in an attempt to achieve emotional regulation (Zilioli et al. 2017). Also jibing with McGarvey is Ruth Lister, who stresses that when people feel that for too long they have been "ignored, dismissed and bullied" (DMG 76), acting out anger or "getting back" at the state, society, or "richer" individuals might also be seen as a way, however problematic, of coping and regaining **agency** (Lister 2021).

Poverty affects the experiences a person is likely to have as well as shaping, probably negatively, the options she has available for interacting with other people. This may result in further social pushback of "the poor" and conversely, contribute to further discrimination. McGarvey refers to an almost absurd incident where for him a moment of perceived discrimination led to deeply felt frustration and resentment. Assigned to psychological counseling for anger management as a teenager, Darren found himself venturing outside Pollok weekly, still a rare thing for him, to meet his counselor. Getting off public transport, his first impression was "an odd feeling of relaxation. People here looked and sounded different in a way that was immediately apparent. … Having taken a few moments to catch my breath, adjusting my eyes to the wonderful technicolour, I remember my first thought being: 'So, this is how people dress when they aren't afraid of being stabbed?'" (DMG 25–26). This eye-opening experience points to what one could call different, and telling, types or "**habitus**"—in other words, to an "embodied" social position, to "the result of the internalization of external structures" (Bourdieu and Wacquant 1992), including social interactions that shape one's appearance and lifestyle (Bourdieu and Wacquant 1992; Bourdieu 1989; Woodward 2018). Yet Darren's sudden insight into the concept of *habitus* is soon overlapped and overshadowed by an encounter with a group of local schoolchildren, who, instinctively, also know about *habitus* and difference: "I immediately sensed they were non-threatening and as they drew closer I overheard them talking. … They were expressive and uninhibited with one another," something almost impossible in Pollok, as other passages of McGarvey's book suggest. Darren feels comfortable: "A part of me wanted so much to walk over and join the conversation, as it seemed like we would probably have had a lot of things in common, but as I passed them they suddenly went quiet. Instantly I knew why: that's what you did when you were walking past something that made you anxious" (DMG 27). The situation had changed in a moment: "these kids seemed to perceive me as a threat"—a disturbing experience he was not at all used to from his everyday life, for in Pollok, young Darren himself was usually the one who felt threatened and attacked by others and he had learned quickly to adapt to the constant threat of violence. "It was a jarring role reversal and I experienced a mix of pride at being feared and resentment at feeling misunderstood" (DMG 27).

One specific aspect of Darren's adaptive behavior consisted of deliberately concealing his interest in words, in the nuances of language and the complexities of its use, especially where the official language was concerned. He hid his longing for proper conversations. These interests, however, led later to involvement in various activist engagements, to a career as a rap performer (professionally known as Loki), and finally to writing *Poverty Safari*. They shaped how McGarvey, looking back in his book, could assess his personal story against the background of society as a whole; but they also led to his heightened attention to the "veiled threats" (Rawlinson 2013) of poverty and eventually to a remarkable change in attitude toward his mother and the narrow life imposed on her by poverty and poverty-related stress. "My mum never got the chance to gain insight into how abnormal her life was; that there was another way of thinking and of being. She never had anything else to compare it to … in my mother's eyes, where I once saw only hatred I now see pain, trauma and a deep frustration at longing to feel connected but not knowing how" (DMG 198–199). It seems plausible that this is something that many people living in poverty and living in disadvantaged areas feel: that they are somehow left behind, that their wish to connect is futile, and that for society as a whole "they count for nothing" (Lister 2015).

Although *situations of poverty* may be quite different—just as a favela shack in São Paolo is definitely something different from a house or flat in Pollok—many *dimensions of the stressful experience of poverty* apply to people living in poverty, regardless of time and place. A group of poverty researchers from the University of Oxford has found in several participatory research projects "that shame is as much an important part of the experience of poverty" in countries such as "China, India, Norway, Pakistan, South Korea or Uganda, as it is in the UK" (ATD Fourth World 2014; see also Walker 2014, 2019). Further participatory research projects confirmed this for other societies all over the world. Findings from the project on "Hidden dimensions of Poverty" (Bray et al. 2019), for example, emphasize the "suffering in body, mind and heart" induced by poverty, not least because of the anxiety "of what others will say on being 'found out' as poor." As we have seen, such feelings can lead to even more avoidance of contact with others "for fear of being judged or shamed" and hence to increased social isolation. "Suffering in body, mind and heart" can be read as an alternative, more figurative wording of "stress" or "poverty-related stress"; indeed, this study calls "stress" one of the core experiences of poverty (Bray et al. 2019).

There is a deeply personal side to the experience of poverty, as the accounts of Carolina Maria de Jesus and Darren McGarvey as well as the voices heard in many reports from participatory research projects convey in much detail. Poverty and sources of poverty-related stress, however, are by no means limited to the personal environment; they are as effectively and closely linked to structures, to the social and political system. The broad research project on the hidden dimensions of poverty conducted by Bray and colleagues, for instance, speaks of "institutional maltreatment" and "social maltreatment" as part of the experience of poverty (Bray et al. 2019). There is a social side of poverty and of the suffering that results from it. There are social, moral, and political aspects of poverty-related stress that need to be addressed; we will do this in the following two chapters.

References

ATD Fourth World (2014) The roles we play. Recognising the contribution of people in poverty. https://atd-ukorg/2014/10/10/the-roles-we-play-recognising-the-contribution-of-people-in-poverty/ Accessed 10 Feb 2021

Attar BK, Guerra NG, Tolan PH (1994) Neighborhood disadvantage, stressful life events, and adjustment in urban elementary-school children. J Clin Child Psychol 23:391–400

Bauman Z (2004) Wasted lives. Modernity and outcasts. Blackwell, Oxford

Bourdieu P (1989) Social space and symbolic power. Sociol Theory 7(1):14–25

Bourdieu P, Wacquant L (1992) An invitation to reflexive sociology. University of Chicago Press, Chicago

Bray R, De Laat M, Godinot X, Ugarte A, Walker R (2019) The Hidden Dimensions of Poverty. International participatory research. Montreuil: Fourth World Publications. https://atd-fourthworld.org/wp-content/uploads/sites/2019/05/Dim_Pauvr_eng_FINAL_July.pdf Accessed 10 February 2021

de Jesus CM (2003) Child of the dark. The diary of Carolina de Jesus. Penguin, London

Edge I, Kagan C, Stewart A (2004) Living poverty: surviving on the edge. Clin Psychol 38(special issue: poverty and social disadvantage):28–31

Fuller-Rowell TE, Evans GW, Ong AD (2012) Poverty and health: the mediating role of perceived discrimination. Psychol Sci 23(7):734–739. https://doi.org/10.1177/0956797612439720

Kaplan SA, Madden VP, Mijanovich T, Purcaro E (2013) The perception of stress and its impact on health in poor communities. J Community Health 38:142–149. https://doi.org/10.1007/s10900-012-9593-5

Levine R (1992) The cautionary tale of Carolina Maria de Jesus. Working paper 178. Kellogg Institute for International Studies. Notre Dame: University of Notre Dame

Lewis O (1959) Five families: Mexican case studies in the culture of poverty. Basic Books, New York

Lister R (2015) 'To count for nothing': poverty beyond the statistics. British Academy lecture read 5 February 2015. J Br Acad 3:139–165. https://doi.org/10.5871/jba/003.139

Lister R (2021) Poverty. Key concepts, 2nd edn. Polity Press, Cambridge

McGarvey D (2018) Poverty safari. Understanding the anger of Britain's underclass. Picador, London

Münzel T, Gori T, Babisch W, Basner M (2014) Cardiovascular effects of environmental noise exposure. Eur Heart J 35:829–836. https://doi.org/10.1093/eurheartj/ehu030

Narayan D (2000) Can anyone hear us? Voices of the poor. Oxford University Press, Oxford

Nicholas MW (1963) Review of "Child of the dark. The diary of Carolina Maria de Jesus". Hisp Am Hist Rev 43(3):448–450

Piña Cabrera L (2010) Calle y escritura como espacio y campo de acción. El testimonio de Carolina María de Jesús, mujer, negra y cartonera. Revista de la Universidad Bolivariana 9(25):487–513

Rawlinson RM (2013) A mind shaped by poverty. Veiled Threads. Regenia M. Rawlinson

Shildrick T, MacDonald R (2013) Poverty talk: how people experiencing poverty deny their poverty and why they blame 'the poor'. Sociol Rev 61:285–303. https://doi.org/10.1111/1467-954X.12018

Sullivan S, Kelli HM, Hammadah M, Topel M, Wilmot K et al (2019) Neighborhood poverty and hemodynamic, neuroendocrine, and immune response to acute stress among patients with coronary artery disease. Psychoneuroendocrinology 100:145–155. https://doi.org/10.1016/j.psyneuen.2018.09.040

Thoits PA (2010) Stress and health: major findings and policy implications. J Health Soc Behav 51(S):S41–S53. https://doi.org/10.1177/0022146510383499

Walker R (2014) The Shame of poverty. With Bantebya-Kyomuhendo G, Chase E, Choudhry S, Gubrium E, Lødemel I, Yongmie JO, Mathew L, Mwiine A, Pellissery S, Ming YAN. Oxford: Oxford University Press

Walker R (2019) Measuring absolute poverty: shame is all you need. In: Gaisbauer HP, Schweiger G, Sedmak C (eds) Absolute poverty in Europe. Interdisciplinary perspectives on a hidden phenomenon. Policy Press, Bristol, pp 97–118

Woodward K (2018) The relevance of Bourdieu's concepts for studying the intersections of poverty, race, and culture. In: Medvetz T, Sallaz JJ (eds) The Oxford Handbook of Pierre Bourdieu. Oxford University Press, New York, NY. https://doi.org/10.1093/oxfordhb/9780199357192.013.29

Zavaleta Reyles D (2007) The ability to go about without shame. A proposal for internationally comparable indicators of shame and humiliation. Oxford Poverty & Human Development Initiative (OPHI). OPHI working papers 03. Oxford: Oxford Department of International Development

Zilioli S, Imami L, Ong AD, Lumley MA, Gruenewald T (2017) Discrimination and anger control as pathways linking socioeconomic disadvantage to allostatic load in midlife. J Psychosom Res 103:83–90. https://doi.org/10.1016/j.jpsychores.2017.10.002

Zimmer U (2013) Nicht von schlechten Eltern. Meine Hartz-IV-Familie. S. Fischer, Frankfurt a. M.

10

Social and Moral Aspects of Stress

As we have seen, stress is a phenomenon that can be read and understood at both an individual and a societal level. This allows for a sociological reading of stress, which we explore in this chapter. In his famous book, *The Sociological Imagination*, Charles Wright Mills describes the sociologist's task as connecting her or his personal **life-world**[1] with the structural realities of the wider framework; this requires what Mills calls "the sociological imagination" as an expression of the vocation of a person doing sociology: "Perhaps the most fruitful distinction with which the sociological imagination works is between 'the personal troubles of milieu' and 'the public issues of social structure'" (Mills 1959). The person "is a social and an historical actor who must be understood, if at all, in close and intricate interplay with social and historical structures" (Mills 1959). This influential invitation to connect individual experiences with a wider social and structural perspective can prove fruitful for a social analysis of stress. For good reasons, Francesco Bottaccioli has called for "stress science as a whole organism science," criticizing a reductionist and mechanist model. He argues for "Psycho-neuro-endocrino-immunology," an interdisciplinary field that considers the human organism as a structured and interconnected unity, in which biological and psychological systems influence each other reciprocally (Bottaccioli and Bottaccioli 2017; Bottaccioli 2014). In this chapter, we combine this concept with the social and sociological dimensions.

An individual person may experience stress, but the experience takes place within a wider framework of structures and systems. To give one example:

---

[1] Glossary terms are bolded at first mention in each chapter.

© Springer Nature Switzerland AG 2021
M. Breitenbach et al., *Stress and Poverty*, https://doi.org/10.1007/978-3-030-77738-8_10

raising a child with special needs is stressful. An honest description of this experience has been provided by Ian Brown in his memoir *The Boy in the Moon*. He describes with fearless candidness the burdens and challenges of parenting in this situation, as well as the beauty and the unexpected gifts. The stress generated in raising a son with severe disabilities runs through the book from the very beginning: "Tonight I wake up in the dark to a steady, motorized noise. Something wrong with the water heater. Nnngah. Nnngah ... But it's not the water heater. It's my boy, Walker, grunting as he punches himself in the head, again and again" (Brown 2009). This is an individual person's experience with another individual person. The situation is one of many stressful situations in their lives. However, there is more to this story than individual interactions: there is a health care system with medical facilities, a social system with support and monitoring services, an educational system with the challenge of providing appropriate support for the child, a legal infrastructure with the charge of regulating state assistance and political and moral matters, including, for example, the question of abortion. All these structural components help to navigate and modify the experience of stress. It is this social perspective that is of interest in this chapter.

On the Sociology of Stress

Without doubt, stress can be reconstructed as a social and even political phenomenon. An important voice in the sociology of stress has been Leonard I. Pearlin, who developed a specific stress model (Aneshensel and Avison 2015). He has followed Mills's advice to narrow the gap between sociological study and everyday life: "It is my intention ... to bring the study of stress closer to the study of ordinary lives. To understand how stress comes about, we do not always have to reach out to the exotic, rare, or eventful. We need only to take a careful look at the structure of everyday experience in the pursuit of everyday life" (quoted in Aneshensel and Avison 2015). He has also made explicit reference to Hans Selye, who has been mentioned many times in this book. Pearlin reconstructs Selye's work, reflecting on the growing field of research on life events: "First, the pioneering work of Hans Selye ... provided an important theoretical foundation for events research. Second, in response to Selye's theoretical inspiration, a method was developed to assess in seemingly simple and objective fashion the magnitude of eventful change experienced by individuals. Third, interest in research into life events was spurred by its early success in showing relationships between the scope of eventful change and various indicators of health" (Pearlin 1989). Pearlin takes as his point of departure the sociologically important dynamics that ensue as a person's well-being, health, and stress levels are affected by the structured arrangements of the person's life and by the repeated experiences caused by these arrangements. There is a social, even a political etiology of ill health, he finds. In his analysis of the sociology of stress, he distinguishes between **stressors**, stress mediators, and stress outcomes (Pearlin 1989). Stressors are factors that cause stress; stress mediators are coping resources that help mitigate the detrimental effects of stressors; and stress outcomes are the manifestations of organismic stress, the effects of the experience of stress. Stressors can be categorized as "primary stressors" (directly related to a stress-causing event and experienced first) and "secondary stressors" (indirectly related to a stress-causing event and occur as a consequence of primary stressors). In addition, stressors occur in different forms. Wheaton and colleagues, for instance, distinguished five major types of stressors: chronic, traumatic, nonevents, daily hassles, and life events (Wheaton et al. 2013). With specific regard to chronic stressors, Wheaton identified seven different subtypes: perceived threat; structural constraints; under-reward; uncertainty; conflict without resolution; overburdening demands; complexity (Wheaton 1997).

One could obviously refine this model by suggesting a more complex analysis; one version of such an analysis was presented in Chap. 7 on stress "**language games**" looking into the nuances of the usage of the term. Here is a slightly more differentiated suggestion for an analysis of the phenomenon of stress:

X experiences S in mode M and in the framework F because of causes C (and reasons R) with the consequences K.

"X" would be the subject of stress (an individual, but possibly also a collective)[2]; S stands for "stress," the alert, strained state that X is in. The "mode M" of stress relates to the question of the way stress is manifested and the way it is imposed on the subject; this is an issue of stress mediators, but also an issue of what could be called "stress filters," filters through which stress is processed. For example, one such important stress filter is gender—for social reasons, men and women may experience stress differently (American Psychological Association 2010; Verma et al. 2011; Mayor 2015)[3]. The

---

[2] The term "stress" is sometimes also applied to institutions as in "stress of financial institutions," or "market stress." There is the well-established OFR Financial Stress Index (OFR FSI), a daily market-based snapshot of stress in global financial markets, constructed from more than 30 financial market variables. Exploring the connection between a biological description of stress and such ways of using the term is an area of further research.

[3] The New York Times described a "stress gap" between men and women on November 14, 2018 (https://www.nytimes.com/2018/11/14/smarter-living/stress-gap-women-men.html, accessed 15 September 2019).

"framework F" is the set of structural and macro-conditions—raising a disabled child in a democratic society is different from raising a disabled child in a dictatorship, for instance, and so a good example of the political framework of stress would be the discussion of the stress associated with raising a disabled child in a developing country (Harriss-White 1996). The "causes C" can be biological in nature, but the experience of stress may have elements of social psychology and other aspects (political, structural, systemic). The "reasons R" are mentioned to provide a perspective that allows us to see the human person as a subject of the stress and stress experience and not merely as a biologically determined being. There are both "causes" and "reasons" that we need to take into account when we try to understand a person's behavior (Taylor 1964, 1985, 1985a). The same applies to the experience of stress—there are identifiable physiological causes, but also relevant (e.g., political) reasons why people experience stress. For example, in certain institutional settings, such as an elite school or a military training facility, the generation of stress is intentional and based on a rationale. There are debates, too, about the optimal stress levels in elite sport settings (Hanton et al. 2009). Stress, like poverty, is often created and produced; it does not simply happen. Inducing stress is a way of governing, exercising control, influencing behavior, and shaping relationships.

Stress leads not only to diverse physiological outcomes, including severe health problems, as discussed in previous chapters, but also to other consequences, such as possible impacts on performance levels, the quality of relationships, or quality-of-life issues. These consequences are quite often social in nature. Stress can also have a "contagion effect," i.e., the effect that the stress on one individual has when it spills over and affects others. If a parent is stressed in her job, it will affect her relationships with her spouse and children.

Stressful experiences have, as Ludwig Wittgenstein would say, an "environment" (Wittgenstein 1967), and this environment is shaped by power dynamics as well as material and structural factors. As we have already seen, there are biological consequences of stress: stressors affect the **autonomic nervous system** and the **HPA axis,** leading to changes in cardiovascular and neuroendocrine measures. But these biological consequences have a "social face." Even though a person is a biological being, she is also a social being whose bodily reactions are conveyed through social contexts and political frameworks. The "biology of stress" may be the same: "whether a person suffers a breakdown from a divorce, an uncompromising boss, or bankruptcy, the breakdown entails the same chemical and hormonal responses regardless of the source of stress" (Au 2017). Yet the "experience of stress" differs depending on the individual situation and social circumstances. There are objective aspects of stressors as well as subjective factors—Dohrenwend emphasizes the objective ones,

Lazarus and Folkman the subjective ones (Dohrenwend 2006; Lazarus and Folkman 1984). According to Au, "job loss can cause stress for an individual, but it may be more prominent if experienced by one who bears the role of a parent. In this instance, the role of parenthood intensifies the stress from job loss with regard to the parent's responsibility to provide for his/her children and family" (Au 2017). In addition to objective and subjective factors, there is also the phenomenon of a "social patterning of stress and coping" (Meyer et al. 2008).

Consequently, one stressor identified by Pearlin in his stress model is what he calls "roles" (Pearlin 1989). **Roles** describe ways of being and acting that cannot be seen in isolation, and are tied to specific social environments and circumstances. Roles are not always a matter of choice for an individual, and it is very likely that a person has more than one role. There can be "role overload" (role demands exceed a person's personal capacities); "interpersonal conflicts within role sets" (problems stemming from interactions between and among persons in complementary roles, such as parent–child, wife–husband, supervisor–worker); "inter-role conflict" (conflicts between different roles a person may have, such as the role of mother and the role of daughter); "role captivity" (a person is stuck unwilling in a role); or "role restructuring" (where stress is generated by the change and transformation of roles) (Pearlin 1989). Roles touch questions of identity and hence can be a source of trouble and disruption. Roles have in this context been mentioned in the World Bank Poverty Study as a source of stress: "Over and over again, in the countries studied women are identified, and identify themselves, as homemakers, the keepers of the family, responsible for the well-being of their children and husbands." The researchers also "relate the entrenched nature of men's identities as breadwinners and decisionmakers even as these roles are undermined and eroded by changing social and economic environments. These socially defined roles of men and women are not only unattainable, they sometimes are in stark contradiction with reality. This is what creates the stress that seems to be endemic in poor households today" (Narayan 2000). Further down we read: "When their authority is challenged men seem to experience stress and exert their right to control the women in their lives through threats and violence. Moreover, this violence, depending on prevailing social norms and structures, may even be naturalized by the victim and perceived as acceptable or normal." Many female respondents felt that men had collapsed under the stresses of poverty (Narayan 2000).

Alain Ehrenberg's study on the sociology of mental health, *The Weariness of the Self*, also confirms the significance of roles. He claims that mental health issues are increasing because people in Western societies are more and more pushed into roles that they do not want, that people don't "own," roles that are not "who they are." It is stressful to be pushed into a role that does not correspond to your personality or value system (Ehrenberg 2010). There is also the phenomenon of "moral stress" (or "moral **distress**"), when "one knows the right thing to do, but institutional constraints make it nearly impossible to pursue the right course of action" (Jameton 1984). This phenomenon is frequently discussed in the healthcare sector, especially in the context of nursing, but also in connection with teaching or work **integrity** (Lützén et al. 2003; Colnerud 2015; Cribb 2011). Pearlin identified chronic strains, such as enduring problems or conflicts, as stressors with a frequent link to role issues: "problems rooted in institutionalized social roles are often enduring, for the activities and the interpersonal relationships they entail are enduring" (Pearlin 1989). Permanent roles and chronic stress can be plausibly connected with each other.

Certain roles are particularly stressful, such as the role of caregivers. "Caregiver stress" can be reconstructed as a consequence of a process that includes socioeconomic characteristics as well as primary stressors (hardships and problems directly related to caregiving) and secondary stressors (either the strains experienced in contexts outside of caregiving or intrapsychic strains involving the diminishment of self-concepts; Pearlin et al. 1990). It is both a sociological (empirical) and a moral (normative) question as to which segment of the population is most often charged with caregiving responsibilities. There is a question of the distributive justice of exposure to stress. In fact, "women are exposed to more stressors associated with the *cost of caring*, such as having greater responsibilities for social network members or experiencing more deaths of friends" (Meyer et al. 2008).

How are high risks of stress socially distributed? This is a matter of **vulnerability**. People have different levels of social support, and they also find themselves at different levels of vulnerability. Paradoxically, quite often people who are in higher need of social support have the lowest **social capital** and, given the dynamics of **social exclusion**, have a hard time trying to attain social integration. This dilemma says something important about vulnerability, as we know that social integration is "associated with *better* physical and mental health regardless of exposure to stress, while measures of social support (usually phrased as general emotional concern or perceived availability of emotional or instrumental support in the face of stress) often buffer the impact of stress, especially on mental health, and only sometimes have main or additive effects on health as well" (House 1987).

Vulnerability can be characterized as the condition of being "susceptible to harm, injury, failure, or misuse" (Formosa 2014). You can be damaged, even destroyed. In turn, violence can be defined as the use of force against persons or things intending to damage or destroy. Vulnerability in this sense is part of the human condition. It means that we cannot reduce life risks—understood as potential impediments to reaching important life goals—to zero. If we take "health" to mean a second-order capability to deal with our first-order capabilities and limits in a way that we can still achieve important life goals (Nordenfelt 1987)[4], then vulnerability threatens health. Here again, we find a vicious circle: stress increases vulnerability, vulnerability increases the probability of stress; stress and vulnerability can be seen as threats to health, and health is a major factor in any account of "quality of life." One issue that will become more and more relevant and burdensome in the future is environmental stress, due to the increased environment-related vulnerability of chronically poor people (Scott 2006). Eric Klinenberg, in his book *Palaces for the People*, discusses the detrimental effects of a 1995 heatwave on different parts of Chicago. It is understood that poor people who live in substandard housing, frequently in areas without backyards, have a hard time protecting themselves from heat waves, but Klinenberg discovered surprisingly vast differences in death rates between two adjacent areas that were comparably (and severely) deprived in demographic terms. In the particularly badly affected area, he found that the neighborhood's deteriorated physical conditions and social infrastructure could explain the significantly higher death rates (Klinenberg 2018). In the less affected neighborhood, a robust social infrastructure was able to buffer stress and vulnerability even in the face of such a catastrophe, keeping death tolls low because people had an eye on each other. As we have seen in previous chapters, in both social and physical terms, a poor neighborhood can very likely see a decline in or even collapse of what Klinenberg calls "social infrastructure," and thus a kind of indifference toward vulnerability that in effect increases vulnerability.

There are some especially vulnerable groups, such as children or the elderly or, in some contexts, single parents or parents of children with special needs. The levels of vulnerability are unequally distributed: "A person is vulnerable to the extent to which she is not in a position to prevent occurrences that would undermine what she takes to be important to her" (Anderson 2014). Some aspects of vulnerabilities could be called "contextual," some could be called "chronic," some are even "pathological." People with more assets, people with more power, people with better health, people with more cultural resources, are less vulnerable. Catriona Mackenzie distinguishes those social

---

[4] In another capability-based understanding of health, health is interpreted as a person's ability to achieve or exercise a cluster of basic human activities (Venkatapuram 2013).

pathologies that create or enhance vulnerabilities from inherent or situational pathologies and introduces the term "pathogenic vulnerability" (Mackenzie et al. 2014). Social pathologies as the causes of pathogenic vulnerabilities undermine the conditions necessary for an environment appropriate for a culture of peacefulness. Social pathologies are forms of social suffering and destructive forms of social development that inhibit human flourishing. One particular aspect of the impact of social pathologies is people's inability to make proper judgments. According to Christopher Zurn's reconstruction of Axel Honneth's approach, social pathologies operate "by means' of second-order disorders, that is, by means of constitutive disconnects between first-order contents and second-order reflexive comprehension of those contents, where those disconnects are pervasive and socially caused" (Zurn 2011).[5] There is a gap between reflection and experience. This analysis views a social pathology in terms of "ideology," i.e., as false beliefs on a first-order level connected with a social inability on a second-order level (an inability to identify the need for reflexivity). In other words, social pathology prevents a person from understanding the mechanisms that caused the condition. An unemployed person, for example, is conditioned to blame herself, thus constructing unemployment as an individual failure, even though there are structural issues at work that the imposed narrative does not take into account. This is a social pathology. Sometimes people accept something morally outrageous (such as the growing levels of inequality within a society and worldwide) as "normal," even though there are many reasons to believe that these dynamics have been caused by socially and morally unhealthy conditions. Social pathologies cause pathogenic vulnerabilities.

Pathogenic vulnerability may lead to chronic and persistent poverty. Chronic conditions change the social game: it is one thing to get short-term emergency support, but it is another matter to receive permanent social support. In addition, support resources may be depleted over time (Aneshensel 1996). It is difficult to sustain social capital in a situation of prolonged social support needs. Sometimes it is a lack of knowledge and information that prevents a person or a family from using social support systems that are in place. This is tragic since networks and relationships protect against toxic stressors (Blair and Raver 2012). This finding has been confirmed in the World Bank's participatory poverty study: "long-term stresses can overwhelm informal support systems. Overuse depletes the capacity of individuals and groups to

---

[5] In his rejoinder in the same volume, Honneth accepts this analysis as exceptionally fruitful (Honneth 2011). A more basic reconstruction of Honneth's concept of social pathologies is provided by Jean-Philippe Deranty (2009).

maintain reciprocal relationships. Kinship and community social networks are resilient, but under times of stress they are less capable of functioning as effective and dependable support systems. During such times the radius of trust and cohesion often narrows to the immediate family, and even family bonds can fracture if pressed too hard" (Narayan 2000).

Before we deepen our understanding of increasing stress levels leading to increased vulnerability by tracing a "moral grammar of events," let us highlight yet another type of social pathology, namely "social maltreatment." For people living in poverty, this means that they are not only "othered" (see Chap. 9.), but that additionally they are often confronted with public behavior that consists of "prejudicial negative judgements, stigma and blame" (Bray et al. 2019; see also Thomas et al. 2020). Here, we may again draw on the definition of vulnerability offered by Anderson, who emphasizes the idea that vulnerability involves both a lack of effective control and an imbalance of power between a person and external forces that influence her (Anderson 2014). Prejudicial negative public opinion and narratives on "the poor" or "underclass" are such external forces contributing to stress and vulnerability. There is a "symbolic violence" to be found in pervasive public discourse on poverty and the assumed lack of "self-responsibility" of "the poor," which even includes blame for poverty-related distress (Thomas et al. 2020). A UK participant in the "Hidden Dimensions of Poverty" study reports that living in poverty "is being treated like cattle, you have no dignity and no identity" (Bray et al. 2019). In the same study, a participant from Bolivia describes social maltreatment as "discrimination because we haven't got any money, we're not well dressed, we haven't studied, … we don't speak properly" (Bray et al. 2019). "To have studied" alludes to the real choices and options a person could have; "to speak properly" points to aspects of development and behavior as well as to having options to fulfill social and cultural norms. This dynamic concerning lack of options relates to findings from previous chapters: poverty and poverty-related chronic stress limit people's real choices and, often from a very early point in a person's life, set pathways for future development. Many of these limitations are signs of structural violence that disempowers people and are beyond an individual's control (Magnuson and Votruba-Drzal 2009; Blair and Raver 2012; Jakovljevic et al. 2016; Treccani 2019). People do not "arrive" in a society; they experience society as if they did not belong. They have to justify their existence or their being in a country (like people who do not master the official language). This unfortunate architecture of social exclusion was succinctly expressed by a US participant in the "Hidden Dimensions of Poverty" study: people usually judge you "by what you have. When you have not much, you are not much. And then, you are not treated like you belong" (Bray et al. 2019). This is the bitter truth about poverty shared by people who know poverty, in particular when they have seen better days and have then fallen into the poverty trap.

A Moral Grammar of Events

Especially vulnerable groups have a hard time preventing life events that increase stress levels and lead to chronic stress. Let us look at one example of an increased level of vulnerability. In a micro-study, the British poverty researcher David Hulme describes a family from Bangladesh over a long period of time (Hulme 2003; Hulme and Moore 2010). This is the story in a nutshell: Maymana and her son Mofizul live in a village about 30 kilometers outside the city of Mymensingh in central Bangladesh. There was a time, in the early 1990s, when this household had five members—Maymana, her husband Hafeez, and their three children (two girls and a boy). At that time Hafeez owned three rickshaws that were an important pillar of the family's livelihood since the rickshaws could be hired out on a daily basis. The family also owned an acre of paddy land, so the household had a reasonably secure income as well as an asset base to fall back on in hard times. In official terms, the family's economic situation would have probably been assessed as a little above the poverty line, although it was an "occasionally poor" household.

At some point in the 1990s, Hafeez began to develop health problems; he had throat pain and coughed a lot. This illness was a stressful life event, without a doubt. Initially, he obtained medicines from a "pharmacist" in the bazaar, almost certainly a person without any formal training or deep expertise. This "treatment" did not make a difference. Then Hafeez visited the nearby government-run health center; he was asked to offer bribes, but even then, the staff did not seem really interested, let alone helpful. So, he went to a "doctor" in a nearby town. It is unclear what "doctor" means in this context, possibly a person with only some medical training and partial studies. This "doctor" prescribed (or rather recommended) expensive medicines. They, too, did not work. The "doctor" had to refer Hafeez to a better-placed colleague in the nearest city, Mymensingh. The costs were piling up and the family had to sell a rickshaw to pay the medical bills. Unfortunately, the condition worsened and more costly measures, X-rays, and further tests were required. Another rickshaw had to be sold. With only one rickshaw left, the weekly household income plummeted, and the family had to cut back on consumption. Hafeez got sicker and sicker.

The first daughter of the family, now of a "marriageable age," was worried about her dowry, so she acquired a kid, fattened it, sold it, and repeated this cycle. In this way she was able to save up for a proper dowry; her younger sister followed her sister's example and saved up for hers. (Male) members of the extended family arranged marriages for the two young women.

By now Hafeez was confined to the house and in bad shape, having lost a lot of weight and gotten weaker. Meanwhile, all the rickshaws had been sold off, the household depended on rice produced from its small plot of land and on Maymana's income from occasional work as a domestic. In 1998, shortly after a stay in the hospital, Hafeez died. His death was another stress-inducing life event: Maymana was in despair with no husband, minimal income, and a disabled child with a weak constitution, Mofizul. Things got even worse. Maymana's father-in-law took control of the household's agricultural plot, and she had to start borrowing, gleaning, and begging for food. Interventions with the village court did not help. Ignoring Bangladesh's official laws, the court ruled against her, and she lost the land.

Maymana and Mofizul received some help from the married daughters, the extended family, neighbors, and the mosque committee. Mofizul tried to find casual employment at a local timber mill, but could not contribute much to the household because of his fragile health as well as his still very young age. The household had fallen into chronic poverty.

Hulme's study reveals life events (illness, death, weddings) that have influenced the stress level of this particular household. We can observe the following factors that added to the stress of the household and contributed to its unnecessary fall into chronic poverty: (1) the failing state, demonstrated by poor basic health services struggling with corruption; lack of access to education and educational opportunities; lack of health insurance; the deficiencies of the legal system that did not support Maymana and her son after the death of her husband; (2) a market for labor, healthcare, and insurance that did not properly include Maymana and Mofizul; the high cost of available health services and insurance packages; (3) the failing society: the mosque committee was charitable, but alms do not allow for planning; the village court cheated Maymana out of her land rights; the household experienced local NGOs as useless; (4) a dysfunctional family: the father-in-law seized Maymana's land; her own father was unable to help; the children were unable to prevent the household from falling into chronic poverty.

A household that is not supported by the state, market, society, or family will not be well equipped to deal with stressful life events. It is part of our human condition that we cannot reduce our risks to zero, but there is the question of the distributive justice of moral luck. Moral luck means that life has spared you the experience of finding yourself in situations that require moral heroism; moral luck describes circumstances when an agent is morally assessed on the basis of actions or personality traits that are formed by factors out of the agent's control—as Thomas Nagel puts it: "it is intuitively plausible that people cannot be morally assessed for what is not their fault, or for what is due to factors beyond their control" (Nagel 1979). Life events such as a stroke in the family with subsequent care responsibilities may lead to chronic strains, but chronic strains themselves (such as poverty, with its effects of diminished quality of sleep and consequently reduced attention levels) may lead to life events such as an accident at work.

There is an important way of offering a social interpretation and understanding of events, as the experiences we have are embedded in social constellations and political frameworks. Even so-called "natural disasters" have "a social face." To give an example, Jon Sobrino reflected upon the devastating earthquake that occurred in El Salvador in 2003. At first glance, an earthquake is a natural disaster and, in this sense, "neutral," as it affects the wealthy and the poor, the saints and the criminals. However, a closer analysis clearly reveals a social dimension to the earthquake: the poor live in the riskiest areas, in the most fragile homes, with the highest probability of collapsing; they live in slums without paved roads, making access for ambulances impossible; they have no options for alternative places to stay, which makes evacuations difficult (Sobrino 2003/2004). Relatedly, using the example of the flooding of New Orleans caused by Hurricane Katrina in 2005, Paul Kadetz has shown how disasters as well as humanitarian crises "challenge existing levels of social capital" and how vulnerability "resulting from poor social capital is inevitably worsened by crises" (Kadetz 2017). There is a "social signature" to the aftermath of disasters such as an earthquake, a hurricane, or a flood. An earthquake may be seen as a "natural disaster," but still affects poor people to a much greater extent than the well-off. The same applies to other disasters; earlier we saw the same dynamics in Klinenberg's analysis of the Chicago heatwave. Life events must be seen through a social perspective in which life events and problems do not exist in isolation; they are part of a web of events and structures.

The study of "the moral grammar of life events" can help us to see that alongside the social dimension of stress there is also a cultural one, one connected to "the meaning of stress" (McLeod 2012). There is no doubt that ideologies influence interpretations of life circumstances, and this leads us to a more explicit discussion of "the politics of stress" in the next chapter.

References

American Psychological Association (2010) Gender and stress. https://www.apa.org/news/press/releases/stress/2010/gender-stress. Accessed 15 Sep 2019

Anderson J (2014) Autonomy and vulnerability intertwined. In: Mackenzie C et al (eds) Vulnerability. OUP, Oxford, pp 134–161

Aneshensel CS (1996) Consequences of psychosocial stress: the universe of stress outcomes. In: Kaplan HB (ed) Psychosocial stress: perspectives on structure, theory, life-course and method. Academic, New York, pp 111–136

Aneshensel C, Avison WR (2015) The stress process: an appreciation of Leonard I. Pearlin. Soc Ment Health 5(2):67–85

Au A (2017) The sociological study of stress: an analysis and critique of the stress process model. Eur J Ment Health 12:53–72

Blair C, Raver CC (2012) Child development in the context of adversity: experiential canalization of brain and behavior. Am Psychol 67:309–318

Bottaccioli F (2014) La fine della grande illusione del riduzionismo in biologia e in medicina. Epistemologia an Italian. J Philos Sci 1:5–21

Bottaccioli F, Bottaccioli AG (2017) Psycho-neuro-endocrino-immunology paradigm and cardiovascular diseases. In: Fioranelli M (ed) Integrative cardiology. Springer, Cham, pp 139–151

Bray R, De Laat M, Godinot X, Ugarte A, Walker R (2019) The hidden dimensions of poverty. International participatory research. Fourth World Publications, Montreuil

Brown I (2009) The boy in the moon: a father's search for his disabled son. St Martin's Press, New York

Colnerud G (2015) Moral stress in teaching practice. Teach Teach Theory Pract 21(3):346–360

Cribb A (2011) Integrity at work: managing routine moral stress in professional roles. Nurs Philos 12:119–127

Deranty JP (2009) Beyond communication: a critical study of Axel Honneth's social philosophy. Brill, Leiden

Dohrenwend BP (2006) Inventorying stressful life events as risk factors for psychopathology: toward resolution of the problem of intracategory variability. Psychol Bull 132:477–495

Ehrenberg A (2010) The weariness of the self: diagnosing the history of depression in the contemporary age. McGill-Queen's University Press, Montreal, CAN

Formosa P (2014) The role of vulnerability in Kantian ethics. In: Mackenzie C et al (eds) Vulnerability. OUP, Oxford, pp 88–109

Hanton S, Thomas O, Mellalieu SD (2009) Management of competitive stress in elite sport. In: Brewer BW (ed) Handbook of sports medicine and science. John Willey, Oxford, pp 30–42

Harriss-White B (1996) The political economy of disability and development with special reference to India. In: United Nations research Institute for Social Development (UNRISD) discussion paper 73. UNRISD, Geneva

Honneth A (2011) Rejoinder. In: Petherbridge D (ed) Axel Honneth: critical essays. Brill, Leiden, pp 391–421

House JS (1987) Social support and social structure. Sociol Forum 2(1):135–146

Hulme D (2003) Thinking 'small' and the understanding of poverty: Maymana and Mofizul's story. IDPM working paper 22. University of Manchester

Hulme D, Moore K (2010) Thinking small and thinking big about poverty: Maymana and Mofizul's story updated. Bangladesh Dev Stud 33(3):69–96

Jakovljevic I, Miller AP, Fitzgerald B (2016) Children's mental health: is poverty the diagnosis? BC Med J 58(8):454–460

Jameton A (1984) Nursing practice: the ethical issues. Prentice-Hall, Englewood Cliffs, NJ

Kadetz P (2017) Social capital in crisis: the complex relationships between community, sociality, inequality and resilience. In: Kapferer E, Gstach I, Koch A, Sedmak C (eds) Rethinking social capital. global perspectives from theory and practice. Cambridge Scholars Publishing, Newcastle upon Tyne, pp 171–188

Klinenberg E (2018) Palaces for the people: how social infrastructure can help fight inequality, polarization, and the decline of civic life. Penguin, New York

Lazarus RS, Folkman S (1984) Stress, appraisal, and coping. Springer, New York, NY

Lützén K, Cronqvist A, Magnusson A, Andersson L (2003) Moral stress: a synthesis of a concept. Nurs Ethics 10(3):312–322

Mackenzie C, Rogers W, Dodds S (2014) Introduction. In: Mackenzie C et al (eds) Vulnerability. OUP, Oxford, pp 1–32

Magnuson K, Votruba-Drzal E (2009) Enduring influences of childhood poverty. Focus 26(2):32–37

Mayor E (2015) Gender roles and traits in stress and health. Front Psychol 6:779

McLeod JD (2012) The meaning of stress: expanding the stress process model. Soc Ment Health 2(3):172–186

Meyer IH, Schwartz S, Frost DM (2008) Social patterning of stress and coping: does disadvantaged social status confer more stress and fewer coping resources? Soc Sci Med 67(3):368–379

Mills CW (1959) The sociological imagination. Oxford University Press, New York

Nagel T (1979) Moral luck. In: Nagel T (ed) Mortal questions. Cambridge University Press, Cambridge, pp 24–38

Narayan D (2000) Can Anyone Hear us? Voices of the poor. Oxford University Press, Oxford

Nordenfelt L (1987) On the nature of health: an action-theoretic account. D. Reidel, Dordrecht

Pearlin LI (1989) The sociological study of stress. J Health Soc Behav 30(3):241–256

Pearlin LI, Mullan JT, Semple SJ, Skaff MM (1990) Caregiving and the stress process: an overview of concepts and their measures. The Gerontologist 30(5):583–594

Scott L (2006) Chronic poverty and the environment: a vulnerability perspective. In: CPRC working paper 62. ODI, London

Sobrino J (2003) Terremoto, terrorismo, barbarie y utopía. UCA Editores, El Salvador. English edition: Sobrino, J. (2004). Where is God? Earthquake, Terrorism, Barbarity, and Hope (trans: Wilde, M.). Maryknoll, NY: Orbis

Taylor C (1964) The explanation of behaviour. Routledge and Kegan Paul, London

Taylor C (1985) Human agency and language. Philosophical papers I. Cambridge University Press, Cambridge

Taylor C (1985a) Philosophy and the human sciences. Philosophical papers II. Cambridge University Press, Cambridge

Thomas F, Wyatt K, Hansford L (2020) The violence of narrative: embodying responsibility for poverty-related stress. Sociol Health Illn 42(5):1123–1138. https://doi.org/10.1111/1467-9566.13084

Treccani G (2019) From structure to behavior: circuit specificity of stress-induced synaptic plasticity in the basolateral amygdala projection neurons. Biol Psychiatry 85(3):E7–E9. https://doi.org/10.1016/j.biopsych.2018.11.017

Venkatapuram S (2013) Health, vital goals, and central human capabilities. Bioethics 27(5):271–279

Verma R, Balhara YPS, Gupta CS (2011) Gender differences in stress response: role of developmental and biological determinants. Ind Psychiatry J 20(1):4–10

Wheaton B (1997) The nature of chronic stress. In: Gottlieb BH (ed) Coping with chronic stress. Plenum, New York, pp 43–73

Wheaton B, Young M, Montazer S, Stuart-Lahman K (2013) Social stress in the twenty-first century. In: Aneshensel CS et al (eds) Handbook of the sociology of mental health, 2nd edn. Springer, New York, pp 299–323

Wittgenstein L (1967) Philosophical investigations. Blackwell, Oxford

Zurn CF (2011) Social pathologies as second-order disorders. In: Petherbridge D (ed) Axel Honneth: critical essays. Brill, Leiden, pp 345–370

11

The Politics of Stress

Stress is *caused*—and so is poverty; political frameworks impact stress levels, and poverty situations as well. Our book has repeatedly highlighted the fact that the word "stress" is used in different ways, by different disciplines, in different contexts. "Stress" can refer to a subjective experience ("I am stressed"), but also to an objectified account ("X experiences stress"). This chapter discusses the political dimension of stress, and we can state right at the beginning that there is a political dimension in the heterogeneous semantics of "stress." Sylvia Tesh has provided an analysis of the 1981 Professional Air Traffic Controllers Organization strike (Tesh 1984); the workers on strike employed the concept of stress, carefully avoiding "subjective claims" by quoting scientific evidence to characterize their working conditions as stressful. By necessity, they had to be selective. "Unfortunately, different experts define stress in strikingly different ways, allowing for a kind of shell game in which the very existence of stress can become problematical. In addition, the stress experts have been unable to offer more than weak data to support the theories that link difficult working conditions to pathological outcomes" (Tesh 1984). This is a fascinating analysis since it challenges the very idea of this book, namely the plausible transferability of a term used in biochemistry to the social sciences, i.e., having a term migrate from the sciences to the social sciences (and then wander off into ordinary language). The point here is not so much the political cost of semantic vagueness, but the political dimension of stress and the facts that (a) the experience of stress is politically relevant (and can offer a basis for labor disputes) and that (b) use of the term "stress" has a political dimension. In this chapter, we will discuss "the politics of stress" with its implications for poverty by looking into the ideas of order and stress distribution.

© Springer Nature Switzerland AG 2021
M. Breitenbach et al., *Stress and Poverty*, https://doi.org/10.1007/978-3-030-77738-8_11

Order and Integrity

Claude Bernard (1813–1878), a French physiologist who is considered the founder of experimental medicine, "was considering that far from being indifferent to the external world, the organism picks up with it a precise and informed relation, in such a way that an equilibrium results from delicate and sustained compensations as it is with a very sensitive balance ... The internal environment corresponds to the fluids and [Bernard] framed the famous sentence 'Constancy and stability of the internal environment is the condition that life should be free and independent'" (Le Moal 2007, referring to Bernard 1878). Here we find the idea of order. Bernard recognized the necessity for organisms to reach some stabilizing independence from the contingencies and changes of the external environment; they attain this stability by cultivating the *milieu intérieur* (inner milieu) or in a more modern language, the non-equilibrium distribution of ions and metabolites across the plasma membrane of the cell. Organisms rely on order to function and survive, e.g., to maintain the unequal distribution of ions across the plasma membrane, which is absolutely necessary for the survival of the cell. There are external and internal factors involved in that order. Order is a stable arrangement of systemic elements that allow for predictability and reduced surveillance costs. In a predictable environment, the effort of monitoring and the sense of the stakes are lower in comparison to a situation where living beings have to be watchful and are at risk of losing their lives. One expression of "environmental order" on the human level with clear laws and hierarchies is the social dimension—i.e., the idea of a social order that allows for peaceful coexistence. The concept of social order refers to stable relational arrangements that allow for the coordination of individual behavior and the cooperation of relevant social players. Here we move into the sphere of politics.

Politics is the art of managing power, especially the power to impose normative expectations. A main task of politics is the establishment and maintenance of a stable social order; social order requires coordination and cooperation. "For social order to arise and be maintained, two separate problems must be overcome. People must be able to coordinate their actions and they must cooperate to attain common goals" (Hechter and Horne 2009). Individual interests and individual **agency**[1] must be subsumed under rules, and common ground in terms of common goals and interests must be established. The negotiation of individuality and communality, with key pillars such as freedom and equality, must be ensured. Social order is about some

---

[1] Glossary terms are bolded at first mention in each chapter.

kind of **homeostasis** too, some kind of equilibrium that balances individual agency and common goals. Tranquility, regularity, reliability, and predictability are features of a social order that reduces the stress of having to adapt to changing environments.

Social order can be challenged or even disrupted because of cultural stressors (Nair et al. 2013) and because of political developments—which are important aspects of "the politics of stress." Political developments can put pressure on individuals and create stressful situations, even stress-determined **life-worlds**. Wars and famines are political events. Wars create war-related stress (Hobfoll et al. 1991, drawing on the example of the Persian Gulf War) and in addition, **post-traumatic stress disorder** or "war in the head" (Laufer et al. 1984, referring to the Vietnam War). Famines are not simply natural disasters, but also the result of political decisions and actions influencing access to information and support systems (Sen and Drèze 1999). Political mismanagement led to China's Great Famine in the late 1950s and early 1960s that resulted in the loss of millions of lives (Jisheng 2012; Meng et al. 2015). Famine, food shortages, and food insecurity equal "food stress" or "food-related stress," because of the loss of equilibrium between needs and resources, the experience of excessive demands, and inadequate response mechanisms and options for agency. "The combination of poor nutrition with the stress that accumulates from being food insecure can contribute to mental health consequences that can last a lifetime. These consequences include, but are not limited to stress, depression, mood disorders, anxiety, substance abuse, stressed relationships, suicidal risk and poor sleep quality" (Weissman 2017). This is another reminder that stress can be a cause of further stress. There is the potential of toxic "stress sustainability" and "reiterative stress." **Toxic stress** is comparable to Selye's understanding of **distress**, or to McEwen's understanding of **allostatic overload**, both previously discussed in this book (Selye 1976; McEwen 2005). Toxic stress undermines the possibility of sustaining a state. The simple point, which can be corroborated by many more examples, is that political decisions on the macro-level can create stressful conditions on the level of communities and individuals. "Loss of freedom" can lead to severe stress—living in a totalitarian regime is stressful, and totalitarian regimes create "tyrannical stress" (Abed 2004).

The disruption of order that causes stress can happen in two ways—either by way of a disruption of the external order or by way of a disruption of the internal order. Both types of disruptions can cause stress if we accept that "stress may be either external with environmental source, or caused by internal perceptions of the individual" (Shahsavarani et al. 2015). Consequently, we need to distinguish between a disruption of the environment and a disruption of the inner landscape. People can be stressed because of a financial loss (a loss of external stability), and they can also be stressed because of a loss of meaning (a loss of inner stability). These two types of disruption are interconnected, since "external factors are not in their essence stressful and/or threatening; yet the individuals' perceptional systems interpret them as such" (Shahsavarani et al. 2015). There is a necessary moment of framing and judging in the act of experiencing, an aspect of categorizing and "experiencing as," i.e., experiencing something as something. Let us take a simple sentence from a biology journal: "Stressors are real or perceived challenges to an organism's ability to meet its real or perceived needs" (Greenberg et al. 2002). This simple statement points to a distinction between "reality-based stress" and "illusion-based stress," and suggests that stress can be caused by misinformation. This is not to say that all "perceived stress" is illusion based, but it points to the need for a proper basis for perceived stress, a *fundamentum in re*. There is always a moment of perception in the experience of stress. That is why, from an epistemological point of view, we can ponder the factor of a necessarily subjective moment in the experience of stress. The subjective dimension of stress is relevant from a poverty studies perspective since the project of "making sense of stress"—an important aspect of coping with stress—also depends on epistemic resources, education, and a set of available categories, all of which poverty situations make more difficult to access.

The dynamics of poverty can also be seen in these two dimensions—in an external dimension of "objective" scarcity and existential pressures, and in an internal dimension of "epistemic aspects," i.e., beliefs, perceptions, judgments. The effects of poverty lead to erosions of both dimensions, since poverty not only puts pressure on issues such as housing, food, and bodily health, but also on issues such as mental health, education, and self-esteem. In a study based on interviews with residents in low-income neighborhoods, Elaine Batty and John Flint explore the connection between neighborhood and self-esteem; the latter is influenced by the former, especially given the pressure and power of comparisons, "from the internalisation of personal critique, self-blame and a sense of not being clever or resourceful enough to manage the consequences of living on a lower income" (Batty and Flint 2010). Neighborhoods are important for an individual's sense of security, "and this could be severely adversely affected by crime and anti-social behavior" (Batty and Flint 2010). Housing is an obvious example of the political dimension of stress. Living in certain neighborhoods can be dangerous and stressful; Jill Leovy has discussed this point in a widely acclaimed report on gang violence in Los Angeles. She describes the permanent stress levels: "Black men who lived in Watts were in constant danger. Those who sold drugs were in more danger" (Leovy 2015). And selling drugs quite often seems to be the most accessible means to livelihood. Gang members live the stress-filled lives of those in constant danger of becoming victims of violent crimes. But the same is true for their families, especially after losing a family member to gang violence: "For many family members, the nightmare begins with experiences most Americans associate only with war: the sudden, violent death of a loved one on the street outside your home ... Immediately after the murders, many of the bereaved describe feeling mechanical and numb, their minds spinning, reflexively pushing agony away" (Leovy 2015).

The safety levels in neighborhoods reflect political decisions and show, once again, the political dimension of stress. Leovy's book refers to a particularly painful source of stress, the moral stress of not being able to live up to well-justified moral expectations, especially one's own. This point has been mentioned before, in Chap. 9, in the context of Carolina Maria de Jesus. Poverty can be seen as a cause for "costly integrity," it is challenging to live a life in accordance with one's own moral standards; sometimes the conditions of poverty are so harsh that they result in a deprivation of affordable access to integrity (not being able to access health care for a sick family member, not being able to feed the children, for example). It is difficult to provide for a family, to sustain friendships, to contribute to the common good, if everyday life is a constant battle or a painful struggle, if resources are scarce and pressures are high. Constant stress levels take their toll, induced by a scarcity of resources and lack of support (**social exclusion**), and this toll challenges the proper functioning of a person in her **role**—as parent, as friend, as community member, for example. One has to be careful, though, with generalizations, since the picture is complex (Cooper 2020). The claim we are making is not a general "poor people are poor parents, poor friends, poor community members," but a more cautious one. Let us introduce the term "**identity labor**" here. Identity labor is the kind of work a person has to do in order to build, maintain, and defend her personal and social identity; in other words, to negotiate her place in the social cosmos. An important aspect of one's identity is the set of "memberships" a person cultivates. These memberships in a company, in a circle of friends, in a political party, in a faith community, in a neighborhood, in a country club, etc., are based on explicit or implicit membership fees. There are certain things expected from a member of X; there is a cost associated with membership. Not living up to these expectations puts one's membership at risk. Paying rent is part of the "membership fee" for living in a particular neighborhood; paying tuition fees and academic success are part of the "membership fee" for being a student at a university; offering financial and emotional support is part of the expected "membership fee" for being a member of a family. Poverty puts pressure on people because of the excessive-ness of demands in comparison to the scarcity of resources, which makes it harder for a person affected by poverty to get her "identity labor" done, to "pay her membership fees." If identity labor fails or even partially fails, people will experience social exclusion, which adds to stress levels, since social sup-port structures affect how people respond to stress and what people perceive as stressful.

If a person faces difficulties in performing her identity labor, she will in all likelihood be confronted with a hiatus between external demands and agency, a hiatus that will deepen social exclusion. Conversely, the stable and socially recognized identity of a person leads to an existential equilibrium whereby a person has negotiated a stable relationship with her environment. This stabil-ity could be characterized by the term "**integrity**." Integrity is not only a crucial moral value, it is also a concept that can meaningfully be used in con-nection with stress: "Stress may be defined as a threat, real or implied, to the psychological or physiological integrity of an individual" (McEwen 1999). As we have seen, Lushchak defined **oxidative stress** as "a situation when steady state **ROS** (**reactive oxygen species**) concentration is transiently or chroni-cally enhanced, disturbing cellular metabolism and its regulation and damag-ing cellular constituents" (Lushchak 2014). Stress threatens integrity; it threatens order. In his approach, Sarafino uses the idea of "disharmony between situational demands and biopsychosocial resources" to characterize stress (Sarafino 2002). The term "disharmony" implies that integrity has been challenged.

Integrity is an important term in moral philosophy. "Integrity" can refer to a person's character or an institution's moral performance, but also to the quality of an ecosystem or any other system, such as a database or an admin-istrative device. The integrity of an entity is threatened by disruption and corruption. "Integrity" is associated with "wholeness," "completeness," and "intactness." The term can be connected to "integration" (putting parts into a whole, preserving unity), to "identity" (maintaining the essence of an entity through changes), and to a moral dimension (having a clear position, living with honesty and uprightness). Insofar as it is related to integration, the term "integrity" can be approached as a result of a unification process, of a process that integrates various parts of a person's life and personality into a harmoni-ous whole. Self-integration points to an intact personality that has ordered her desires and her needs well. Here again we can make use of the idea of an order with a set structure and a hierarchy of elements. Harry Frankfurt used the term "wholeheartedness" to refer to a well-organized hierarchy of desires (Frankfurt 1988, especially Chapter 12). This order is based on a hierarchy of commitments, and the integrity of a person can also be seen as founded on the identification, acceptance, and realization of commitments. Identity-conferring commitments shape a person's profile at an existential level and organize a person's agency as a coming together of different types of interests. As Bernard Williams put it, "unless I am propelled forward by the conatus of desire, project and interest, it is unclear why I should go on at all" (Williams 1981).

The concept of commitments is relevant for our topic of stress and poverty: poverty reduces the possibility of entering and honoring commitments due to the unpredictability and fragility of life and a restricted planning horizon; a bank loan is based on a commitment to repay the loan, a commitment that people affected by poverty can often not make. The same applies to other commitments, such as employment commitments. To illustrate this point, Jacqueline Novogratz has described her attempts as an American to build a bakery in Rwanda, with the help of a local women's group. She is frustrated by the lack of reliability of her coworkers, who do not show up on time and cannot honor their commitments. A woman tells her, "our lives have many obligations attached to them. We have funerals and weddings and births and so many commitments" (Novogratz 2009). Honoring a set of commitments B inevitably means dishonoring a set of commitments A. The integration has not happened and cannot happen in a stable way, given the excess of demands. Commitments—and the sense of a constant "commitment struggle," combined with difficulties in entering and honoring commitments—are an important lens through which to think about poverty.

The same is true of stress. Stress is also affected by commitments, since they establish a sense of order and provide a motivation to close any gap between the current situation and the level of commitment. In this sense, stress as a *movens,* a moving force, is enabling and promoting but in a potentially destructive dynamic also eroding and damaging a person's integrity and commitment structure.

In order to maintain integrity, adjustments have to be made. There is a price for this, which has been called "**allostatic load**," referring to "a gradual process of wear and tear on the body" (McEwen 1999). Stress and **stress responses** have an impact on the organism, and this impact can lead to more stress. There is the possibility of vicious circles, whereby stress leads to unhealthy (even self-damaging and self-destructive) behavior, which leads to more stress, which leads to further unhealthy responses. Stress is the result of change and the response to this change; loss of balance and search for new balance; threat to integrity and efforts toward defending integrity. Shahsavarani and his coauthors have characterized stress as any effect of change in surrounding environments on a living being, which results in a disruption of the internal balance (homeostasis) of that living being (Shahsavarani et al. 2013). Poverty is particularly prone to changes in the environment, through phenomena like job insecurity, housing insecurity, family instability, and food insecurity. As we have seen, the increased **vulnerability** levels of poverty are deeply connected to structural issues and political frameworks. This "politics of stress" can also be seen in the distribution of stress.

Distribution of Stress

Political frameworks have a major impact on the mechanisms of the distribution of goods and burdens. In fact, distributive justice is one of the major goals of politics. The unequal distribution of goods can be identified as a key question in political philosophy—how to deal with inequality? Social contract theories from Hobbes's "Leviathan" to Rawls's "Theory of Justice" have framed and debated the question in this way: what are appropriate ways to justify and manage inequality?

Inequality can also be observed in the case of stress. The distribution of stress is uneven—and this uneven distribution refers to both the positive and the negative aspects of stress. Stress as a physiological stimulus connected to human-environment interactions can be positive in terms of moving people to act in order to adjust to the environment; stress serves as a *movens*, a moving force that influences an entity's adaptation to environmental changes and challenges that lead to actions and reactions. As has been stated before, being exposed to a stimulating environment is an important aspect of brain development; living in a world characterized by stimulating change and options can be appealing and attractive. The weight of boredom and the pressure of limited options reduce this experience. Poverty is a deprivation of attractive stimulations; due to limited relational and occupational options, people affected by poverty suffer from monotony, narrowness, restrictions, and lack of positive stress. "**Eustress**" (see Chap. 2) is a desirable good, not easily accessible to people living in poverty. It has been argued for good reason that a certain type of stress is important for learning and stress-related growth (Rudland et al. 2020). For instance, challenges in a learning environment can be stressors, but important stressors can facilitate learning and enhance cognitive development.

Stress is unhealthy and even dangerous if there is a significant imbalance between demands on the individual and the individual's ability to meet these demands. Unhealthy stress can be understood "in terms of general physiological and psychological reactions that provoke adversarial mental or physical health conditions when a person's adaptive capabilities are overextended" (Babatunde 2013). In a stressful situation, a person reaches the limits of adequate functionality. There is also an uneven distribution of unhealthy stress; those who are socially excluded are confronted with various additional sources of stress, many of which are chronic.

There is also the dimension of the politics of gender equality; for example, pressures and stress levels experienced by women during the pandemic have a political dimension. Sadly, this is not new. During the Ebola crisis in Africa, we could observe the same gender gaps in care and the same gender gaps in decision-making. Writing about gender dimensions of the Ebola outbreak in Nigeria, Olufunmilayo I. Fawole and coauthors state: "Women were exposed occupationally and domestically due to their care giving roles. In health facilities, they were directly involved in the care or encountered persons who had been in contact with persons with Ebola. In the homes, they were at the forefront of nursing the sick. There is the need to ensure women have access to information, services and personal protective equipment to enable them protect themselves from infection. Education and engagement of women is crucial to protect women from infection and for prompt outbreak containment" (Fawole et al. 2016). We can safely assume that different stress levels go with these additional pressures and risks. Obviously, gender dimensions of stress appear not only in times of great illness. There are, of course, multiple other examples of gender discrimination piling on top of issues of poverty: low-income workers are burdened further by the need to arrange childcare, traditionally relegated to women; the sad reality that sexual harassment is difficult to deal with in situations where women must hold on to the job, the phenomenon of "missing fathers" leaving women with care-giving and stressful financial responsibilities…

The distribution of justice reflects social inequality; in other words, there is no "stress justice." In the case of stress, inequality discussed at a societal level in philosophical terms can be translated into physiological phenomena at the individual level, which can then be described in biochemical terms. This is another way of saying that there are social determinants of stress, probably an amalgam of interconnected sociocultural and economic factors, as a study from Sri Lanka has shown (Senanayake et al. 2020).

As we have already observed, there is no "stress justice." Poor people are disproportionately exposed to unhealthy stress and disproportionately deprived of eustress. This idea of "stress justice" basically means the fair distribution of access to eustress and the fair distribution of unhealthy stress levels. A theory of stress justice would take into account the link between social injustice and stress injustice. The justification of social inequality has limits if one accepts the dignity of each person and the equality of this dignity. If we further accept equality in the sphere of democratic citizenship, there are even clearer limits as to the justifiability of the unequal distribution of stress within a society. The world we have built shows stress injustice.

This lack of stress justice can be observed in institutional settings and on a societal level; within institutions, we can refer to "occupational stress" and its correlation to "lack of agency." The profession of air traffic controllers mentioned earlier is an interesting point of reference for understanding stress in institutional settings. In a report for the International Labor Organization on occupational stress in air traffic control, Giovanni Costa (1996) lists the following sources of stress for air traffic control personnel: demand (number of aircraft under control, peak traffic hours, extraneous traffic, unforeseeable events), operating procedures (time pressure, having to bend the rules, feeling of loss of control, fear of the consequences of errors), working times (unbroken duty periods, shift and night work), working tools (equipment in terms of their limitations and reliability), work environment (lighting, noise, microclimate, bad posture, rest and canteen facilities), and work organization (role ambiguity, relations with supervisors and colleagues, lack of control over work process, salary, public opinion). Even though this analysis may be dated, the list is eye-opening: there are physical aspects, such as equipment, and individual psychological aspects, such as feelings of loss of control and fear, and social aspects, such as work climate and relationships.

Institutions can influence the physical as well as the social aspects of the work environment. Stressors with regard to occupational stress are often connected to job control (Mostert et al. 2008). A study from Iran identified inadequate pay, inequality at work, excessive workload, poor recognition, time pressure, job insecurity, and lack of management support as key sources of occupational stress (Mosadeghrad 2014). Freedom and recognition emerge as major factors; these are scarce goods in low-income jobs. Job strain is significantly higher in jobs that combine low decision latitude with high workload demands, as in the case of health care workers (Landsbergis 1988). Landsbergis offers some examples of occupational stress: "a 49-year-old bus driver, who commutes 1 hour for a 12-hour split shift, driving through the congested streets of a large city, is trying to keep his hypertension under control. … A 56-year-old hotel housekeeper, who works without rest breaks to complete the many rooms she has to clean daily, is experiencing back and shoulder pain. A 39-year-old temporary worker for a major software company, who works long hours yet gets fewer holidays, vacation days, and pay raises than his salaried coworkers, is experiencing headaches, irritability, and difficulty sleeping"

(Landsbergis et al. 2017). These examples show the idea that the risk of significant levels of occupational stress is especially high in low-income and low-prestige jobs (which does *not* mean low-responsibility jobs, thinking of the bus driver or the health care worker or indeed the cleaner in time of COVID-19). Institutional stress levels depend on the level of privilege and power within the institution, as confirmed by the famous Whitehall studies (Marmot et al. 1978, 1991). "Job control" is a major issue in stress-related symptoms. Low levels of control in the work environment have been associated with an increased risk of future **coronary heart disease** among government employees (Bosma et al. 1997). In addition, social status matters for health (Marmot 2005). In academic institutions, tenure is the most important stabilizing factor at the faculty level and corresponds to stress levels that are much higher in untenured faculty members (Reevy and Deason 2014; for an interesting case study, see Carr 2014). The key factor here is "contingency" which expresses a lack of possibility for safe planning and lack of security, factors associated with poverty. "Contingency" means that so many things *could* happen—and these possibilities of disruption challenge the possibility of planning, of entering, and honoring commitments.

It is evident that the distribution of stress within an institution does not follow patterns of social justice. Individuals work within institutions, and institutions are embedded in society, which can lead to the phenomenon of "stress transfer." "Stress transfer" means that pressure is produced in a larger unit A and pushed down to the smaller unit B. This pressure is translated into stress at the level of individuals. A school in the United States, for instance, is required to administer standardized tests, which puts pressure on teachers and students and creates stressful situations. In order not to lose their accreditation, schools have to abide by these rules, even though the testing culture may go against the ethos of the school. This is a case of stress transfer. We can also observe "shifting of stress" in a hierarchically structured institution where stress produced on a higher level (e.g., financial pressures on hospitals) is translated into stress on a lower level (nurses have to work under much more pressure). The infamous Mid Staffordshire hospital scandal in the United Kingdom, resulting in hundreds of deaths, is an example of this dynamic (Francis 2013; see also Holmes 2013; Halligan 2013; Smith and Chambers 2019).

On a societal level, it is evident that some people are exposed to sources of stress for structural reasons, not because they necessarily did anything wrong. Iris Marion Young, in discussing structural injustice, tells the story of Sandy: Sandy loses her apartment close to her job because the building was bought by a real estate developer who intends to convert the cheap apartments into expensive condominiums, a process typically described as "**gentrification**." Paying the price for gentrification, Sandy needs to relocate and can only find affordable housing far from work. If she wants to keep her job, she has to buy a car since the bus service would not be feasible. "Sandy sees no other option but to take the apartment, and then faces one final hurdle: she needs to deposit three months' rent to secure the apartment. However, she has used all her savings for a down payment on the car. She cannot rent the apartment and, having learned that this is a typical landlord policy, she now faces the prospect of homelessness" (Young 2011). Due to no fault of her own, Sandy has lost her life basis. She is pushed into a stressful spiral of difficult decisions and awful choices; the stress she has to deal with is political. Similar stories of victims of gentrification can be easily found (see, for example, Moskowitz 2018).

Let us offer another example: Matthew Desmond, the author of *Evicted* and other influential academic publications on evictions (Desmond 2017), wrote an article for the *New Yorker* about a particular eviction story in Milwaukee, Wisconsin. It tells the story of Arleen Beale, a single mother of a 13-year-old and a 5-year-old boy. Arleen had experienced a series of evictions. One eviction, described at the beginning of the article, says a lot about the vicious circle of poverty and stress. On a particular January day in a notably snowy winter, Arleen's son and his cousin entertained themselves by throwing snow at the passing cars. "One jerked to a stop, and a man jumped out, chasing the boys to Arleen's apartment, where he broke down the door with a few kicks. When the landlord found out about the property damage, she decided to evict" (Desmond 2016). It is very clear here that it is stressful for a parent who has to make ends meet to look after her children; it is stressful for a single parent to know that she cannot control the risk levels and the social acceptability of her child's behavior. It is stressful to have your door kicked in, especially if this damage is done to someone else's property and puts you at the risk of losing your home. It is stressful to interact with an angry landlord. Finally, it is stressful, incredibly stressful, to be evicted. Evictions can lead to posttraumatic stress disorder (Robles-Ortega et al. 2017). In the case of Arleen, who has had her share of eviction experiences, housing insecurity has shaped most of her adult life. This has undeniable effects on her neurological situation and her stress levels in the eviction experience narrated in Desmond's article. As McEwen (1999) puts it: "The brain interprets what is stressful on the basis of past experience of the individual."

Evictions cause stress and affect the health of the people involved (Desmond and Tolbert Kimbro 2015; Pevalin 2009; for a remarkable case study, see also Qvarfordt Eisenstein 2016). It is stressful to go through a conflict that ends with housing instability; it is stressful to be forced to pack up belongings and leave a place that you have called "home." It is stressful not to know where to go next, especially if you have family responsibilities. The process itself is stressful since it involves authorities and force: "The landlord would summon the sheriff, who would arrive with a gun, a team of movers, and a judge's order saying that her house was no longer hers." And then there were two options: "truck" (all belongings loaded into a truck and checked into bonded storage) or "curb" (all belongings piled up on the sidewalk). The former is only available for those who can pay. The latter means that all your belongings, including the few delicate and precious things you own, are exposed to the elements and to theft and robbery. Sometimes the eviction process is "informal," which has its own kind of stress because of the less rule-driven and more arbitrary nature of the process in which the (former) tenants are at the mercy of the house owner.

Unhealthy stress, as mentioned before, creates a situation where the level of demands exceeds an individual's ability to deal with the demands—and the experience of eviction is clearly a case of unhealthy stress, since losing a home means losing an identity, a refuge, a livelihood basis, an existential foundation. It is an existential space, as Judith Sixsmith has shown in an important article (Sixsmith 1986). Losing a home affects the deep layers of a person. A home offers emotional security, and lack of a home equals physical insecurity, which touches upon a basic need. The causes of eviction are stressful and the process of eviction is stressful, but so are the consequences. It is stressful to live in a homeless shelter with its rules, changing clients, dearth of privacy and protected spaces, and its temporary and transitional nature.

Evictions are stressful, and evictions are not the fate of high-income people; housing instability is an expression of financial and social instability. There is also a political dimension to evictions. Let us go back to an episode in Arleen's life, as reconstructed in Desmond's *New Yorker* article. After staying in a homeless shelter with her boys, Arleen found a house on Milwaukee's North side. "There was often no running water, and Jori had to bucket out what was in the toilet. But Arleen loved that the rent was only five hundred and twenty-five dollars a month, and that the house was set apart from others on the block. 'It was quiet,' she remembered. 'It was my favorite place.' After a few weeks, the city found the house 'unfit for human habitation.' Arleen moved into a drab apartment complex deeper in the inner city, on Atkinson Avenue, which she soon learned was a haven for drug dealers. She feared for her boys, especially Jori, who was goofy and slack-shouldered and would talk to anyone" (Desmond 2016). Eventually, she found an alternative place and while at first things seemed to turn out well, this time the next eviction happened because of her falling behind on her rental payments. Why did she fall behind?

She had to contribute to her sister's funeral costs, and her welfare check was reduced because she missed an appointment. And why did she miss the appointment? Because the reminder note was sent to a previous address, from which Arleen had already been evicted. In other words, simplifying it slightly, Arleen experienced housing insecurity because of a previous experience of housing insecurity. Her story reflects the profound shortage of public housing, a political dimension that transcends an individual's agency.

This dimension beyond the individual's sphere of influence refers to the politics of stress. Stress is caused, created. So is poverty. Poverty is caused and created. The politics of poverty and the politics of stress are connected.

In a previous chapter, we quoted Ruth Keil's observation that we do not use the word "stress" in cases of large-scale violence such as ethnic cleansing. Keil contrasts such experiences to quite different ones, such as "the irritation about clerical work." She explains, "the amount of stress a person feels about form-filling is to some extent under their own control. There is an implication, therefore, that stress is an individual event, rather than political or social, and also that it can be controlled or mitigated or decreased, and an industry has sprung up to do exactly this" (Keil 2004). These astute observations point to the dynamics of individualizing systemic failures. People blame themselves for things that go wrong, even though the real problems happen at a systemic level. We described these dynamics earlier in this book as an expression of a social pathology, and can find an illustration in Debra E. Meyerson's ethnographic study on the stress experienced by social workers employed in medical institutions: "Social workers in the acute-care hospitals tended to interpret burnout as a disease of the individual. They blamed individuals for not properly coping with this fate, or they described burnout as a personal character flaw" (Meyerson 1994). Stress levels are windows into societies and institutions. People who experience stress can also be seen as seismographs. Stress can be linked to the "stressed individual who is at once damaged by society and maladjusted to society" (Abbott 1990). This perspective can lead to competing interpretations of stress as an individual problem or as a social phenomenon (Meyerson 1994). A cultural theory of stress reconstructs stress as a symbol that refers to a culturally anchored duality—a duality of the individual as both "agent" and "product," as both acting subject and shaped object (Barley and Knight 1992). The distribution of stress is a result of lack of choice and a matter of being victimized. There can be no doubt that the politics of stress and the politics of poverty are interconnected.

Since politics is translated to the individual level by way of institutions, we can finally ask the question: how do we build stress-sensitive institutions and stress-sensitive societies?

The question is an open one, but it is clear that a well-functioning welfare state can prevent, reduce, and mitigate poverty and poverty-related stress (Lee and Koo 2016). Legislation and institutions can effectively buffer the stress of poverty; in a number of countries (e.g., in Europe), the legislator protects people from potentially devastating existential risks by ensuring access to public health care, access to a pension system, access to free higher education. The development of the welfare state is obviously the result of many political debates and decisions many of which are relevant for the "politics of stress."

References

Abbott A (1990) Positivism and interpretation in sociology: lessons for sociologists from the history of stress research. Sociol Forum 5(3):435–458. https://doi.org/10.1007/BF01115095

Abed RT (2004) Tyranny and mental health. Br Med Bull 72(1):1–13. https://doi.org/10.1093/bmb/ldh037

Babatunde A (2013) Occupational stress: a review on Conceptualisations, causes and cure. Econom Insights – Trends Chall 3(II):73–80

Barley SR, Knight DB (1992) Toward a cultural theory of stress complaints. In: Staw BM, Cummings LL (eds) Research in organizational behavior, vol 14. JAI Press, Greenwich, CT, pp S.1–S48

Batty E, Flint J (2010) Self-esteem, comparative poverty and neighborhoods. Research paper no. 7. Centre for Regional Economics and Social Research; Sheffield Hallam University, Sheffield

Bernard C (1878) Leçons sur les phénomènes de la vie communs aux animaux et aux végétaux. Librairie J.B. Baillière et fils, Paris

Bosma H, Marmot MG, Hemingway H, Nicholson AC, Brunner E, Stansfeld SA (1997) Low job control and risk of coronary heart disease in Whitehall II (prospective cohort) study. BMJ 314:558. https://doi.org/10.1136/bmj.314.7080.558

Carr AR (2014) Stress levels in tenure-track and recently tenured faculty members in selected institutions of higher education in Northeast Tennessee. Dissertation, East Tennessee State University

Cooper K (2020) Are poor parents poor parents? The relationship between poverty and parenting among mothers in the UK. Sociology. https://doi.org/10.1177/0038038520939397

Costa G (1996) Occupational stress and stress prevention in air traffic control. Working paper CONDFT/WP.6/1995. International Labor Organization, Geneva

Desmond M (2016, February 1) Forced Out: for many poor Americans, eviction never ends. The New Yorker. https://www.newyorker.com/magazine/2016/02/08/forced-out

Desmond M (2017) Evicted: poverty and profit in the American City. Random House, New York

Desmond M, Tolbert Kimbro R (2015) Eviction's fallout: housing, hardship, and health. Soc Forces 94(1):295–324. https://doi.org/10.1093/sf/sov044

Fawole OI, Bamiselu OF, Adewuyi PA, Nguku PM (2016) Gender dimensions to the Ebola outbreak in Nigeria. Ann Afr Med 15(1):7–13. https://doi.org/10.4103/1596-3519.172554

Francis R (2013) Report of the mid Staffordshire NHS foundation trust public inquiry. Executive summary. https://assets.publishing.service.gov.uk/government/uploads/system/uploads/attachment_data/file/279124/0947.pdf

Frankfurt HG (1988) Identification and wholeheartedness. In: Frankfurt H (ed) The Importance of what we care about. Cambridge UP, Cambridge, pp S.159–S.176

Greenberg N, Carr JA, Summers CH (2002) Causes and consequences of stress. Integr Comp Biol 42(3):508–516. https://doi.org/10.1093/icb/42.3.508

Halligan A (2013) The Francis report: what you permit, you promote. J R Soc Med 106(4):116–117. https://doi.org/10.1177/2F0141076813484109

Hechter M, Horne C (2009) Theories of social order. Stanford UP, Stanford

Hobfoll SE, Spielberger CD, Breznitz S, Figley C, Folkman S, Lepper-Green B, Meichenbaum D, Milgram NA, Sandler I, Sarason I, van der Kolk B (1991) War-related stress: addressing the stress of war and other traumatic events. Am Psychol 46(8):848–855. https://doi.org/10.1037//0003-066X.46.8.848

Holmes D (2013) Mid Staffordshire scandal highlights NHS cultural crisis. Lancet 381(9866):521–522. https://doi.org/10.1016/S0140-6736(13)60264-0

Jisheng Y (2012) Tombstone: the great Chinese famine, 1958–1962. Farrar, Straus and Giroux, New York

Keil RMK (2004) Coping and stress: a conceptual analysis. J Adv Nurs 45(6):659–665. https://doi.org/10.1046/j.1365-2648.2003.02955.x

Landsbergis PA (1988) Occupational stress faced by health care workers: a test of the job demands-control model. J Organ Behav 9(3):217–239. https://doi.org/10.1002/job.4030090303

Landsbergis PA, Dobson M, LaMontagne AD, Choi BK, Schnall P, Baker DB (2017) Occupational stress. In: Levy BS, Wegman DH, Baron SL, Sokas RK (eds) Occupational and environmental health, 7th edn. OUP, Oxford, pp S.325–S.344. https://doi.org/10.1093/oso/9780190662677.003.0017

Laufer RS, Gallops MS, Frey-Wouters E (1984) War stress and trauma: the Vietnam veteran experience. J Health Soc Behav 25(1):65–85. https://doi.org/10.2307/2136705

Le Moal M (2007) Historical approach and evolution of the stress concept: a personal account. Psychoneuroendocrinology 32(1):3–9. https://doi.org/10.1016/j.psyneuen.2007.03.019

Lee C-S, Koo I-H (2016) The welfare states and poverty. In: Brady D, Burton LM (eds) The Oxford handbook of the social science of poverty. OUP, Oxford. https://doi.org/10.1093/oxfordhb/9780199914050.013.32

Leovy J (2015) Ghettoside: a true story of murder in America. Spiegel and Grau, New York

Lushchak VI (2014) Free radicals, reactive oxygen species, oxidative stress and its classification. Chem Biol Interact 224:164–175. https://doi.org/10.1016/j. cbi.2014.10.016

Marmot M (2005) The status syndrome: how social standing affects our health and longevity. Holt Paperbacks, New York, NY

Marmot MG, Rose G, Shipley M, Hamilton PJS (1978) Employment grade and coronary heart disease in British civil servants. J Epidemiol Community Health 32:244–249. https://doi.org/10.1136/jech.32.4.244

Marmot MG, Smith GD, Stansfeld S, Patel C, North D, Head J, White I, Brunner E, Feeney A (1991) Health inequalities among British civil servants: the Whitehall II study. Lancet 337(8754):1387–1393. https://doi.org/10.1016/0140-6736(91)93068-k

McEwen B (1999) Stress. In: Wilson RA, Keil F (eds) The MIT encyclopedia of the cognitive sciences. MIT Press, Cambridge, MA

McEwen BS (2005) Stressed or stressed out: what is the difference? J Psychiatry Neurosci 30(5):315–318

Meng X, Quian N, Yared P (2015) The institutional causes of China's great famine, 1959–1961. Rev Econ Stud 82(4):1568–1611

Meyerson DE (1994) Interpretations of stress in institutions: the cultural production of ambiguity and burnout. Adm Sci Q 39(4):628–653. https://doi.org/10.2307/2393774

Mosadeghrad AM (2014) Occupational stress and its consequences: implications for health policy and management. Leadersh Health Serv 27(3):224–239. https://doi.org/10.1108/LHS-07-2013-0032

Moskowitz P (2018) How to kill a city. Gentrification, inequality, and the fight for the neighborhood. Bold Type Books, New York

Mostert FF, Rothmann S, Mostert K, Nell K (2008) Outcomes of occupational stress in a higher education institution. South Afr Bus Rev 12(3):102–127

Nair RL, White RMB, Roosa MW, Zeiders KH (2013) Cultural stressors and mental health symptoms among Mexican Americans: a prospective study examining the impact of the family and neighborhood context. J Youth Adolesc 42(10):1611–1623. https://doi.org/10.1007/s10964-012-9834-z

Novogratz J (2009) The blue sweater. Bridging the gap between rich and poor in an interconnected world. Rodale, New York

Pevalin DJ (2009) Housing repossessions, evictions and common mental illness in the UK: results from a household panel study. J Epidemiol Community Health 63(11):949–951. https://doi.org/10.1136/jech.2008.083477

Qvarfordt Eisenstein C (2016) Depression in the aftermath of eviction. A one-year follow-up study of a disruptive housing life event. Sociology Department: University of Stockholm

Reevy GM, Deason G (2014) Predictors of depression, stress, and anxiety among non-tenure track faculty. Front Psychol 5:701. https://doi.org/10.3389/fpsyg.2014.00701

Robles-Ortega H, Guerra P, Gonzáles-Usera I, Mata-Martín JL, Fernández-Santaella MC, Vila J, Bolívar-Muñoz J, Bernal-Solano M, Mateo-Rodríguez I, Daponte-Codina A (2017) Post-traumatic stress disorder symptomatology in people affected by home eviction in Spain. Span J Psychol 20(e57):1–8. https://doi.org/10.1017/sjp.2017.56

Rudland JR, Golding C, Wilkinson TJ (2020) The stress paradox: how stress can be good for learning. Med Educ 54:40–45. https://doi.org/10.1111/medu.13830

Sarafino EP (2002) Health psychology: biopsychosocial interactions, 4th edn. John Wiley & Sons, New York

Selye H (1976) Forty years of stress research: principal remaining problems and misconceptions. CMA J 115:53–56

Sen A, Drèze J (1999) The Amartya Sen and Jean Drèze omnibus: poverty and famines; hunger and public action; India: economic development and social opportunity. Oxford University Press, Oxford

Senanayake B, Wickramasinghe SI, Edirippulige S, Arambepola C (2020) Stress and its social determinants – a qualitative study reflecting the perceptions of a select small Group of the Public in Sri Lanka. Indian J Psychol Med 42(1):69–79. https://doi.org/10.4103/2FIJPSYM.IJPSYM_482_18

Shahsavarani AM, Ashayeri H, Lotfian M, Sattari K (2013) The effects of stress on visual selective attention: the moderating role of personality factors. J Am Sci 9(6s):1–16

Shahsavarani AM, Azad Marz Abadi E, Hakimi Kalkhoran M (2015) Stress: facts and theories through literature review. Int J Med Rev 2(2):230–241

Sixsmith J (1986) The meaning of home: an exploratory environmental experience. J Environ Psychol 6(4):281–298. https://doi.org/10.1016/S0272-4944(86)80002-0

Smith J, Chambers N (2019) Mid Staffordshire: a case study of failed governance and leadership? Polit Q 90(2):194–201. https://doi.org/10.1111/1467-923X.12698

Tesh S (1984) The politics of stress: the case of air traffic control. Int J Health Serv 14(4):569–587. https://doi.org/10.2190/JH2E-F62P-WMX8-7NQF

Weissman SK (2017) Comparing the physical and psychological effects of food security and food insecurity. Thesis eastern Michigan university. Eastern Michigan University, Ypsilanti, Michigan

Williams B (1981) Moral luck: philosophical papers 1973–1980. Cambridge University Press, Cambridge

Young IM (2011) Responsibility for justice. Oxford University Press, Oxford

12

Responding to Stress and the Value of Resilience

Stress is an adverse experience that cannot be ignored and needs attending to. There has to be an appropriate stress response to stress. In fact, we could say that life is an ongoing and creative response to stress, a process of negotiating one's relationship with the environment and the pressures it puts on an organism or person. In the biochemical model, as cited previously, we have seen with Lushchak how **oxidative stress**[1] disturbs cellular metabolism and damages the cellular constituents of the organism (Lushchak 2014). In the sociological model, stress has also been described as the result of a dissonance between given conditions and an individual's resources (Pearlin and Bierman 2013). **Stressors** challenge a person's adaptive capabilities and impose behavioral adjustments, given the idea of a certain order that allows for functioning and flourishing. Stressors call for a reaction, a response. Defense and recovery mechanisms and strategies are necessary to counteract the detrimental and sometimes deadly consequences of stress.

The last chapter ended with the question, how do we build stress-sensitive institutions and stress-sensitive societies? This is a question that moves us into the realm of response and coping mechanisms. One prominent concept that expresses the flexibility to absorb shocks and pressures and the ability to function or even flourish in spite of adverse conditions is the concept of **resilience**. After having explored stress and its connection to poverty, we want to discuss resilience as an important response mechanism to stress in general and to poverty-related stress in particular.

---

[1] Glossary terms are bolded at first mention in each chapter.

© Springer Nature Switzerland AG 2021
M. Breitenbach et al., *Stress and Poverty*, https://doi.org/10.1007/978-3-030-77738-8_12

The Concept of Resilience

Resilience is a word, like stress, that is used quite frequently, especially in the context of crises, and there is an intuitive understanding of what it means. However, defining resilience precisely and scientifically correctly is difficult. Resilience is often presented as an important way of responding to different types of adversity. A classic definition, coined by ecologist Crawford Stanley Holling, describes resilience as "a measure of the persistence of systems and of their ability to absorb change and disturbance and still maintain the same relationships between populations or state variables" (Holling 1973; see also the collections of articles by Gunderson and Holling 2002 and Gunderson and Pritchard 2002). Resilience is a way of dealing in a nondestructive way with pressure, shocks, and **vulnerability**. The literature explicitly connects "resilience" to "stress." Stress has been recognized as a source of psychopathology, making the understanding of resilience and the identification of its biological basis as well as the understanding of resistance to pathological outcomes of stress relevant at an individual level, but also at a societal one (Cabib et al. 2012). Rutter characterizes resilience as "the reduced vulnerability to risk experiences, the overcoming of a stress or adversity, or a relatively good outcome despite risk experiences" (Rutter 2012). Similarly, in a review article by Wu and coauthors, "Resilience is the capacity and dynamic process of adaptively overcoming stress and adversity while maintaining normal psychological and physical functioning" (Wu et al. 2013). The key terms at stake here are obviously "adaptation," "adversity," and "normal functioning." It is worth remembering that these terms are also prominent in stress research; "adaptation," for example, is the central term in Hans Selye's groundbreaking model of the **General Adaptation Syndrome** (see Chap. 2).

Resilience has been described as the response to shocks and stresses, leading to a distinction between stress-based resilience and shock-specific resilience. According to the *Concise Oxford English Dictionary,* synonyms include "the ability to recoil, rebound, resume shape and size after stretching or compression." This list clearly encompasses but is also much broader than just "resistance." More specifically, referring to usage in the medical literature, the American Heritage Dictionary defines resilience as "the ability to recover quickly from illness, change, or misfortune; buoyancy."

Etymologically, resilience is connected to Latin *resiliens*; with *salire* meaning "to jump, to leap," the present participle of *resilire* ("to rebound, recoil") means to "leap back." Intuitively, we grasp that resilience means the ability to safely return from a state of severe stress to normality, both in terms of physiological parameters (such as, for instance, cellular oxidative stress) and in terms of psychological equanimity and interaction with peers. Prolonged chronic stress or singular events of major life disruptions (already discussed in other parts of this book) induce an **allostatic load** (McEwen 1998), meaning a change in the set point of a physiological dynamic equilibrium (see below) with sometimes-drastic long-term effects, such as **posttraumatic stress disorder** (**PTSD**) (Krystal and Neumeister 2009; Hoge et al. 2007).

There is no consensus regarding an operational definition of resilience, but these details may not even matter if there is agreement that resilience is positive adaptation in the face of adversity, whereby "positive" implies that the adaptation process ensures functioning and functioning well (Herrman et al. 2011).

If we take resilience to be a coping ability, we can distinguish three types of resilience: (1) minimizing the risk of experiencing adverse situations; (2) denying or escaping adversity; (3) facing adversity. The first option is not an option for particularly vulnerable people, who usually lack choices. People with choices can often walk away from stressful situations: they can leave a stressful relationship; they can leave a stressful work environment; they can leave a stressful country. The second option refers to unhealthy ways of dealing with adversity and stress, such as alcohol, drugs, and escapism. The World Bank Study we cited earlier, for instance, has identified this unhealthy coping mechanism as significant: "Alcohol is frequently used to manage and alleviate stress and has a strongly negative impact on household members" (Narayan 2000). The third option is key to making proper use of the term "resilience." Resilience means coping and flourishing either "in spite of" or even "because of" adversity. Exposure to traumatic events may even help build resilience in terms of the "inoculating effect" of stressful experiences (Khoshaba and Maddi 1999; Solomon et al. 2007). Likewise, some adversity can actually help to strengthen and foster resilience (Seery 2011); here we are reminded of the concept of "**hormesis**," a favorable response to a small dose of a potentially toxic drug (see Chap. 1). In this sense, the coping mechanisms of poor people are especially relevant to a deeper understanding of stress resilience, since involuntary poverty has to be understood as an adverse condition. We are convinced that resilience research can benefit a great deal from studying the coping mechanisms of poor households and poor communities more closely.

Let us return once again to the example of Maymana and her family in Bangladesh (introduced in Chap. 10). This was the story of a family falling into severe and chronic poverty as the result of several personal shocks and blows as well as various societal and structural failures (Hulme 2003). Yet the researchers exploring Maymana's household encountered a family that was not paralyzed by poverty but proved to be resourceful and proactive without giving up hope (Hulme and Moore 2010). As described earlier, when the family's situation worsened, the two daughters found ways of generating their own income and thus were able to save themselves for their dowries, allowing them to get married (Hulme 2003). But in addition, Maymana and her son Mofizul, who had seemed stuck in enduring poverty, never gave up. Notwithstanding ill health, physical impairment, and lack of education, they thought strategically about how to make a living and grabbed every opportunity; Mofizul managed to maintain his determination and optimism about the future, which helped him at work, despite his several limitations (Hulme and Moore 2010). Against the odds, Maymana and Mofizul eventually managed to change course and improve their circumstances significantly. These are impressive examples of perseverance and resilience. Those who have to live with permanent and deep adversities can teach us a great deal about what resilience as "functioning in spite of adversity" can mean in practice.

Resilience Research

Emmy Werner, a US psychologist and pioneer in resilience research, studied the development of 698 children born on the island of Kauai, Hawaii, in 1955. Her results show that children from similar backgrounds developed differently, with varying degrees of "success": amazingly and intriguingly, some managed to prosper and flourish while others in the same conditions did not (Werner and Smith 1982, 1992, 2001). Werner identifies protective properties which boost resilience in children (Werner 1992, 1996); these are closely linked with social attitudes and an inner frame of mind strongly influenced by temperament and disposition—a "cheerful soul" will manage better. In addition, Werner particularly emphasizes good relationships, social integration, communication networks, self-esteem, and being able to think in terms of long-term goals and not immediate rewards. How we cope (or not) says a lot about our level of *inner* vulnerability, which will be heightened by neuropsychological deficits, chronic illness, high levels of distractibility, low cognitive dexterity, and weak self-regulatory capacities, and in addition, the *external* risk factors we are faced with, such as enduring low **socioeconomic status**, living in (and being locked into) chronic poverty in poor housing facilities, and experiencing permanent family conflict compounded by the bad health of family members. But the main deficiencies children face are parents with addictive habits and family life in social isolation, thus leading to chronic stress. We have seen in previous chapters the likely detrimental effects of poverty-related stress on a person's development, in particular during childhood.

Beginning with Werner's work, resilience was linked to the pillars of vulnerability and protection. These two aspects still play a major role today. In a definition that comes out of research on child abuse, resilience refers to the "protective and compensatory factors" which can protect the abused child while growing up (Egle et al. 2016). Resilience can be strengthened—but also made less necessary—through proper protection of individuals and through the presence of protective factors in their lives. Vulnerability can be reduced through a well-functioning welfare state, as mentioned at the end of the Chap. 11, and protective factors such as community involvement or meaningful relationships with a person outside of one's family can be strengthened.

It is still an open question why certain people are resilient and others are not. But there is no doubt that people react and respond very differently to adversity. "While susceptible individuals poorly adapt to stressors and express inappropriate responses that can become persistent states of stress, resilient individuals can perceive adversity as minimally threatening and develop adaptive physiological and psychological responses" (Franklin et al. 2012). Even though there are both threatening and protective factors that change the likelihood of resilient responses, we are far from being able to make predictions. The frequently used binary distinction ("either a person is resilient or she is not") does not seem to capture the complexity of realities. For example, a young child's **autonomic nervous system** "can develop either an adaptive stress response that is responsive to everyday challenges or a dysregulated stress response that contributes to or augments the negative impact of adverse life experiences" (Alkon et al. 2012). Rather than working with such binary codes, we might want to embrace a concept of "selective resilience," in which a person may be resilient vis-à-vis life events of type A, but not vis-à-vis life events of type B. A person can cope with her own illness, for example, but is not able to deal with the illness of a close family member. Some people are able to deal with major shocks of significant life disruptions, but not with minor stressors in everyday life.

In any case, we can work with the understanding that resilience is the ability to cope with unfavorable situations; advanced or chronic stress levels constitute particularly unfavorable situations or adverse circumstances. Early on, Werner discussed resilience at the individual level, but since then resilience research has progressed, moving from a more individual to a more political understanding. We can identify distinct phases in the development of resilience research (Layne et al. 2007; O'Dougherty Wright and Masten 2005; Masten 2007). A first phase can be seen in the research on childhood, with the main insight that a protected environment is necessary to lay a solid foundation for building future resilience. Long-term observational studies define a second phase, concentrating on those contextual aspects that accompany biological and sociocultural factors. In a third phase, institutional and political attitudes and aspects of intervention and prevention are addressed (Masten and Obradovic 2006; Masten and Powell 2003). A fourth phase of research sees resilience as a transboundary, multidisciplinary venture.

These four phases work with resilience on different levels. Resilience can be linked to personal biological factors and to systemic environmental factors (Herrman et al. 2011). We can distinguish different levels and types of resilience: (1) the individual, (2) family structures, (3) community and social environments, (4) workplaces and institutions, and (5) the state. "Resilience" can be the property of material as explored by material sciences, of cells and organisms, of individuals, families, communities, institutions, and even of nation-states. In the twenty-first century, with its growing awareness of global interconnectedness and global environmental threats, we could even cultivate the idea of a perhaps dangerously limited "resilience" as a property of the planet. The basic idea—turning or returning to a state of functioning in the face of adversity—can be applied to all these different levels of resilience. The point of this book is to encourage a dialogue between the natural sciences, social sciences, and ethics, discussing the levels of cell, individual, and society with regard to stress and also to the stress response and stress resilience. Resilience can be studied at the cellular level. In fact, it turns out that the mechanisms of resilience that can be studied at the cellular level, as in our investigation into the understanding of stress, can help us to understand resilience at the level of the psychology of the individual and even at the level of groups of individuals. An experiment performed with yeast cells from the laboratory of one of us (Breitenbach[2]), based on the yeast oxidative stress model briefly introduced in Chap. 1, illustrates the dynamics of resilience very well. We will explain the experiment, shown in Fig. 12.1, in simple terminology, with some further experimental details included in the figure's legend.

Figure 12.1a shows yeast cells growing on a synthetic complete medium in the late logarithmic phase. On the left side, the cells are shown by fluorescence microscopy with staining of the **mitochondria**; on the right side, the same cells are shown in phase-contrast microscopy. The cell mitochondria form a continuous network extending from the mother cells into the daughter cells. This is the ground state or normal unstressed appearance of mitochondria that

---

[2] The experiments were performed and documented by Mark Rinnerthaler.

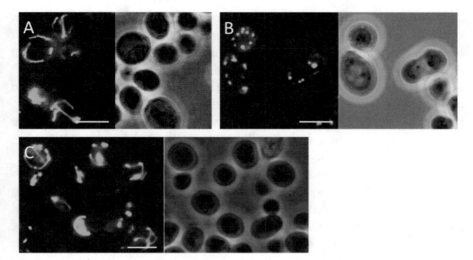

Fig. 12.1 Yeast mitochondria under oxidative stress and after return to normal conditions. Resilience of yeast cells to oxidative stress. This figure illustrates a haploid yeast strain, which expresses the ACO1-GFP construct from a plasmid under control by the MET25 promoter. The construct codes for a fluorescent version of the mitochondrial protein aconitase. The growth medium is synthetic, completely lacking uracil so it can select for the presence of the ACO1-GFP-carrying plasmid. In each sample, fluorescence microscopy appears on the left and phase-contrast microscopy on the right. The length bar indicates 10 μm in each sample. Part **a**: Cells from late exponential growth. Part **b**: Cells from the same culture viewed 90 min after the addition of hydrogen peroxide (1 mM final concentration). Part **c**: Cells from the same culture growing without hydrogen peroxide 75 min after washing and replacing with fresh medium. The morphology of the mitochondria at the end of the experiment shows that the cells have completely recovered, indicating resilience under conditions of oxidative stress

are actively respiring and performing the whole basic array of metabolism, which is necessary for growth.

In Fig. 12.1b, the yeast culture was put under stress by exposing it to 1 mM hydrogen peroxide, which exerts strong but non-lethal oxidative stress on the cells. Consequently, the culture underwent a transitory growth arrest after which it restarted growth, but at a slower pace. The picture was taken 90 min after the cells had resumed growth. Both mother and daughter cells now show many small punctate mitochondria. The same cells are again shown in fluorescence and phase-contrast microscopy. We consider mitochondrial morphology as a useful and easily observable proxy for mitochondrial physiology and respiration.

In Fig. 12.1c, the cells were released from stress exposure, which means they were washed and resuspended in a fresh synthetic complete medium. Again, a transitory growth arrest was observed. After 75 min, growth was resumed, and the cells again showed a mitochondrial network extending from the mother into the daughter cells. During these 75 min, the metabolism of the cells underwent a large transient change, affecting the mitochondria but also the whole cellular system.

In this simple model system, what are the characteristics of resilience, and how can we assess them? Evidently, these cells showed resilience under changing oxidative stress conditions. The figure shows that the cells under stress can adapt and reach a new metabolic allostatic equilibrium, but in addition can dynamically readjust when the growth conditions return to normal.

Resilience in this model system consists of two characteristics: (1) the ability to survive a large amount of stress or in other words, resistance to stress; and (2) the ability to return dynamically to the normal metabolic ground state. It is the second of these two traits that is truly constitutive of resilience. We suggest equating the redox metabolism (and other types of metabolism) of cells that survive and grow under oxidative stress with the allostatic load. The new homeostatic situation may be enough to survive, but not enough to permit all physiological functions that we know from studying the ground state. Another homeostatic equilibrium is reached, but with a different set-point from the ground state homeostatic equilibrium, and therefore, with a different overall metabolic state. To quantify cellular resilience, we suggest a combined measure of the maximum amount of oxidative stress that can be survived, and the time necessary to return to the ground state.

It is easy to imagine how this simple cellular model can be compared to the metabolic changes in neurons of the **hippocampus** (Han and Nestler 2017) under conditions of chronic stress and under the influence of singular catastrophic life events. The parallel between these two stressed cell types—yeast cells in the presence of hydrogen peroxide and neurons of the hippocampus in a psychologically stressed person—lies in the fact that in both cases oxidative stress occurs in the cells, and in both cases, the oxidative stress of the cells can be overcome and reset to normal after a certain time free from stress. (In the case of neurons this has already been discussed in Chap. 4.)

One of the unanswered questions of stress research concerns the stress and stress response factors that in one person lead to severe disease, such as PTSD, and in another person can be overcome. The latter ability we call "resilience under adverse conditions." In terms of psychology, we can describe factors that help a person to become resilient (as described in this chapter); however, in terms of genetics and biochemistry, at the present state of research, it is nearly impossible to define a molecular mechanism of resilience.

We hope that the results shown in Fig. 12.1 provide a concrete and material definition of resilience that is direct and easily comprehensible and can also be correlated with the biochemical results of stress physiology research—and perhaps in the future, with psychological and sociological scales of resilience described in the literature.

Strengthening Resilience

Resilience has been characterized as the ability to return to a state of functioning. Resilience also refers to the capacity to maintain stable functioning and to undergo adaptation in the face of significant adversity (Fletcher and Sarkar 2013; García Secades et al. 2014). Resilience can be strengthened, and it can even be learned, as for example, a study of responses to occupational stress has shown (Chitra and Karunanidhi 2018). Experience of moderate stress can strengthen resilience to future stressful events (Neff and Broady 2011).

From a philosophical perspective, informed by social science, resilience can be strengthened by three simple pillars: social sense, sense of control, sense of direction (Sedmak 2013, especially Chapter 2).

Social sense highlights a person's interest in and connection with others. A study by Anne Broussard and colleagues, capturing the experience of 12 single mothers raising their children under adverse circumstances, identified "volunteering" with other low-income individuals as a coping and stress relief strategy (Broussard et al. 2012). The World Bank study we have mentioned several times identified social support structures as mitigators of poverty-related stress, stating: "societal bonds can help to stabilize communities and ease the psychological stresses of poverty" (Narayan 2000). Drawing on observations from Mexico, the report also mentions "the paradox that while indigenous communities of Oaxaca have the least materially, they are happy and less fearful than nonindigenous poor people because they have a range of 'traditional communitarian institutions which provide them support in times of need'" (Narayan 2000). Social support systems can strengthen stress resilience in a variety of contexts, e.g., in academic settings (Wilks 2008). Social sense is about both *reaching out* and *being reached out to*. Social support has tangible physiological effects; low social support has been associated with physiological and neuroendocrine indices of heightened stress reactivity (Ozbay et al. 2007).

Sense of control highlights a sense of **agency** and choice even when facing constraints and adversity. A cognitive reappraisal of a situation is relevant for stress resilience, and emotion regulation has been recognized as an important protective factor to show resilience in the face of stress (Troy and Mauss 2011). If we think back to the analysis of adverse childhood experiences that often have a significant negative impact on a person's emotion regulation, we see the morally relevant (since injustice-consolidating) paradox that those who are forced into having to be resilient are, at the same time, deprived of important sources of resilience. It has been rightly observed that "one of the consequences of long-term stress exposure is to disrupt processes involved in successful adaptation to environmental threat" (Baratta et al. 2013). One major response to poverty, then, is "freedom," understood as access to agency (Sen 1999), and a sense of control is expressed in the ability to exercise responsible agency. Research shows that resilient human beings have a sense of agency, personal responsibility, and growth (O'Connell Higgins 1994). An example, knitting those two factors together, can be found in the work of Boris Cyrulnik. A French psychiatrist and physician born to émigré Jews in 1937, Cyrulnik lost his parents in Auschwitz and was not treated well by his foster parents; he worked—as a seven-year-old boy!—with the resistance movement. Thus, his existential background for talking about resilience is deep and rich. Cyrulnik characterizes resilience as "a mesh," not "a substance." He uses the image of a "sweater," knit out of many strands of yarn. Similarly, we are invited to knit ourselves sweaters, using the people and things we meet in our emotional and social environments. These experiences, then, are the "wool" that contributes to our emotional clothing (Cyrulnik 2004; Cyrulnik 2009). Resilience is a process, a continuous process of knitting and adding all the various threads and wools we come across as we work—or knit—our way through life. Cyrulnik depicts resilience as a form of anti-fatalism, resisting quirks of fate. People who can actively contribute to their own lives, who do not feel like victims of overpowering and irreversible circumstances, have a better chance of acting resiliently.

A *sense of direction* refers to beliefs and a value system about life projects and priorities. It refers to a sense of purpose, a sense of values. Wu and coauthors observed that "the existence of a moral compass or an internal belief system guiding values and ethics is commonly shared among resilient individuals," referring to a study on depression and stress resilience (Wu et al. 2013; Southwick et al. 2005). Viktor Frankl's well-known insights into the role of "attitude" and a sense of purpose also illustrate the connection between resilience and a sense of direction, a sense about where life should lead and what it is about (Frankl 1984). Building resilience is also a matter of the appropriate framing of life events. Cyrulnik tells a story he attributes to the French poet Charles Péguy: "On his way to Chartres, Péguy saw a man breaking stones with a big sledge hammer by the side of the road. His face was a picture of misery and his gestures were full of anger. Péguy stopped and asked: 'What are you doing, monsieur?' 'You can see what I'm doing,' the man replied, 'this stupid, painful job is all I could find.' A little further on, Péguy saw another man. He was breaking rocks, too, but his face was calm and his gestures were harmonious. 'What are you doing, monsieur?' asked Péguy. 'Oh, I'm making a living. It's hard work, but at least I'm out of doors.' Further on, a third stonebreaker radiated happiness. He smiled as he put down his hammer and looked at the fragments of stones with pleasure. 'What are you doing?' asked Péguy. 'Building a cathedral'" (Cyrulnik 2009a).

The story illustrates the point of being able to attribute weight and meaning to what you do. This sense of direction can be strengthened by social factors (such as identity-giving communities) but dwells especially in a person's inner worldview. These sources of resilience that stem from a person's set of beliefs, hopes, ideas, and values can be called "**epistemic resilience**" (Sedmak 2017, especially Chapter 2.5). This point about epistemic resilience has been succinctly expressed in a thought often ascribed to Friedrich Nietzsche: "Those who have a why to live for can bear almost any how."

To sum up, resilience can be strengthened by accessing three important factors: social sense, sense of control, and sense of direction. These "pillars" can serve as important ways of coping with stress. Unfortunately, these three sources are threatened and weakened in situations of poverty. Firstly, poverty as **social exclusion** undermines **social capital** and social networks; it deprives people of easy access to institutions and communities. It weakens the social sense. The romantic idea of poor communities helping each other is quite often not the reality—we have seen in Carolina Maria de Jesus' as well as Darren McGarvey's testimonies that mistrust and violence can be disturbing side-effects of poverty and can affect poor communities. Second, poverty undermines a person's sense of control due to the limited agency and the restricted set of options available for people affected by poverty. The so-called "capability approach" defines poverty as capability deprivation, a deprivation of real opportunities to act (Hick 2014). Poverty leads to the erosion of a person's choices, restricting "the sphere of agency" and diminishing grounds for **self-efficacy**. The sense of control is weakened. Third, poverty is quite often connected to a lack of access to education and to those sources that constitute epistemic resilience. It may be easier to develop a sense of purpose if one feels strongly about one's agency; it may be easier to deal with external pressures if one is able to cultivate an inner life, and if one can access an inner world of ideas, as a number of examples in this book (Carolina Maria de Jesus, Darren McGarvey, Boris Cyrulnik) have illustrated. Of course, nothing is automatic; poor people can have a deep sense of purpose and educational resources do not automatically lead to a rich "inner life." But poverty, clearly, is an impediment to the ability to make plans and to have long-term goals. As he described, Frankl survived the concentration camp thanks to his long-term perspective and hope grounded in his previous experience as a medical doctor in Vienna. It would be much more difficult to develop such a perspective without previous reference points or role models. That is to say that poverty, with its constant struggle to make ends meet, makes it difficult to develop a long-term perspective that would be helpful for this "sense of direction."

Concluding Remarks

Let us return at the end of this chapter to the moral aspect of resilience. We have seen once again the tragic dynamics of the paradox that those in greatest need of stress-coping mechanisms are systematically deprived of the most plausible sources for these mechanisms. This is not to say, of course, that people affected by poverty should be reduced to passive victims. It is exciting (and probably under-researched) what might be learned from poor people who deal with shocks and stresses day in, day out. Poor people who live their lives with **integrity** in spite of scarce resources undoubtedly have a great deal to teach us about stress resilience. Such people show agency, even "mastery."

Mastery "refers to the individual perception of ability to handle stress … A common variable affiliated with mastery is socioeconomic status, where the higher a person's socioeconomic status is, the more education and occupational background and prospect he/she would have, and consequently the higher his/her sense of mastery" (Au 2017). This is clearly another manifestation of the **Matthew effect**, which in a nutshell can be expressed as, "those who have will be given more." Those with socioeconomic privileges, which reduce the probability of chronic stressors in the first place, have higher coping potential than those in significantly more disadvantageous (and needy) situations. Social stress theory offers a framework for explaining health disparities, making use of a sociological paradigm that views social conditions as a cause of stress for socially disadvantaged people or groups (Dressler et al. 2005; Aneshensel et al. 1991). "Mastery" lessens the effect of stressors by sophisticated cognitive possibilities (such as perceptual or epistemic minimization of the stressor), thus contributing to and building from a basis of self-esteem and a belief of superiority vis-à-vis external developments. Once again, a sense of control reduces the impact of stressors significantly (Au 2017), and a sense of control is a scarce resource for people living in poverty.

There is a twofold injustice involved in the experience and accessibility of resilience: first, some people are resilient and others are not—there is the issue of the distributive justice of access to resilience resources. Second, some people have to be resilient or otherwise they would not be able to live their lives—there is the issue of the distributive justice of the necessity to cultivate resilience. Carolina Maria de Jesus, for example, had no choice but to be resilient; otherwise, she would have starved or ended her life by suicide. Given the differing stress levels of households, some households have no choice but to be resilient, and the pressure to be resilient is also unequally distributed.

This twofold injustice should be changed. We have discussed political aspects of stress and poverty, and the same can be done about resilience. There is a politics of resilience. We hope that we have been able to show that the link between stress and poverty is so strong and recognized within political frameworks that something needs to be done about it—and that something can be done about it. Political conditions can cause and perpetuate stress.

So, the question then becomes how we can build stress-sensitive and resilient societies under real-life conditions. We thought about this question at the end of the last chapter. How can we mitigate or avoid the stress of poverty? We referred to a health insurance system, an old-age pension system, and equal access to education—in other words, to a properly functioning welfare state. In the light of the poverty-related pressures that transcend an individual person's or family's capacity, turning to a systemic safety net seems to be the most realistic and most sustainable solution.

References

Alkon A, Wolff B, Boyce T (2012) Poverty, stress, and autonomic reactivity. In: King R, Maholmes V (eds) The Oxford Handbook of poverty and child development (chapter 12). Oxford University Press, Oxford. https://doi.org/10.1093/oxfordhb/9780199769100.013.0012

Aneshensel CS, Rutter CM, Lachenbruch PA (1991) Social structure, stress, and mental health: competing conceptual and analytic models. Am Sociol Rev 56(2):166–178. https://doi.org/10.2307/2095777

Au A (2017) The sociological study of stress: an analysis and critique of the stress process model. Eur J Mental Health 12(1):53–72. https://doi.org/10.5708/EJMH.12.2017.1.4

Baratta MV, Rozeske RR, Maier SF (2013) Understanding stress resilience. Front Behav Neurosci 7:158. https://doi.org/10.3389/fnbeh.2013.00158

Broussard CA, Joseph AL, Thompson M (2012) Stressors and coping strategies used by single mothers living in poverty. J Women Soc Work 27(2):190–204. https://doi.org/10.1177/0886109912443884

Cabib S, Campus P, Colelli V (2012) Learning to cope with stress: psychobiological mechanisms of stress resilience. Rev Neurosci 23(5–6):659–672. https://doi.org/10.1515/revneuro-2012-0080

Chitra T, Karunanidhi S (2018) The impact of resilience training on occupational stress, resilience, job satisfaction, and psychological Well-being of female police officer. J Police Crim Psychol 3:116. https://doi.org/10.1007/s11896-018-9294-9

Cyrulnik B (2004) Parler d'amour au bord du gouffre. Odile Jacob, Paris

Cyrulnik B (2009) Resilience: how your inner strength can set you free from the past. Jeremy P. Tarcher/Penguin, London

Cyrulnik B (2009a) Talking of love. How to overcome trauma and remake your life story. Penguin, London

Dressler WW, Oths KS, Gravlee CC (2005) Race and ethnicity in public health research: models to explain health disparities. Annu Rev Anthropol 34:231–252. https://doi.org/10.1146/annurev.anthro.34.081804.120505

Egle UT, Franz M, Joraschky P, Lampe A, Seiffge-Krenke I, Cierpka M (2016) Gesundheitliche Langzeitfolgen psychosozialer Belastungen in der Kindheit – ein Update. Bundesgesundheitsbl Gesundheitsforsch Gesundheitsschutz 59:1247–1254. https://doi.org/10.1007/s00103-016-2421-9

Fletcher D, Sarkar M (2013) Psychological resilience: a review and critique of definitions, concepts, and theory. Eur Psychol 18(1):12–23. https://doi.org/10.1027/1016-9040/a000124

Frankl VE (1984) Man's search for meaning: an introduction to Logotherapy. Touchstone, New York

Franklin TB, Saab BJ, Mansuy I (2012) Neural mechanisms of stress. Resilience and vulnerability. Neuron 75(5):747–761. https://doi.org/10.1016/j.neuron.2012.08.016

García Secades X, Molinero O, Ruíz Barquín R, Salguero A, de la Vega R, Márquez S (2014) La resiliencia en el deporte: fundamentos teóricos, instrumentos de evaluación y revisión de la literatura. Cuadernos de Psicología del Deporte 14(3):83–98. https://doi.org/10.4321/S1578/84232014000300010

Gunderson LH, Holling CS (eds) (2002) Panarchy: understanding transformations in human and natural systems. Island Press, Washington, DC

Gunderson LH, Pritchard L (eds) (2002) Resilience and the behavior of large-scale systems. Island Press, Washington, DC

Han M-H, Nestler EJ (2017) Neural substrates of depression and resilience. Neurotherapeutics 14:677–686. https://doi.org/10.1007/s13311-017-0527-x

Herrman H, Stewart DE, Diaz-Granados N, Berger EL, Jackson B, Yuen T (2011) What is resilience? Can J Psychiatry 56(5):258–265. https://doi.org/10.1177/070674371105600504

Hick R (2014) Poverty as capability deprivation: Conceptualising and measuring poverty in contemporary Europe. Eur J Sociol 55(3):295–323. https://doi.org/10.1017/S0003975614000150

Hoge EA, Austin ED, Pollack MH (2007) Resilience: research evidence and conceptual considerations for posttraumatic stress disorder. Depress Anxiety 24(2):139–152. https://doi.org/10.1002/da.20175

Holling CS (1973) Resilience and stability of ecological systems. Annu Rev Ecol Syst 4:1–23. https://doi.org/10.1146/annurev.es.04.110173.000245

Hulme D (2003) Thinking 'small' and the understanding of poverty: Maymana and Mofizul's story. In: IDPM working paper 22. University of Manchester, Manchester

Hulme D, Moore K (2010) Thinking small and thinking big about poverty: Maymana and Mofizul's story updated. Bangladesh Dev Stud 33(3):69–96

Khoshaba DM, Maddi SR (1999) Early experiences in hardiness development. Consult Psychol J 51(2):106–116. https://doi.org/10.1037/1061-4087.51.2.106

Krystal JH, Neumeister A (2009) Noradrenergic and serotonergic mechanisms in the neurobiology of posttraumatic stress disorder and resilience. Brain Res 1293:13–23. https://doi.org/10.1016/j.brainres.2009.03.044

Layne CM, Warren JS, Watson PJ, Shalev AY (2007) Risk, vulnerability, resistance and resilience: toward an integrative conceptualization of posttraumatic adaption. In: Friedman MJ, Keane TM, Resick PA (eds) Handbook of PTSD: science and practice. The Guilford Press, New York, pp 497–520

Lushchak VI (2014) Free radicals, reactive oxygen species, oxidative stress and its classification. Chem Biol Interact 224:164–175. https://doi.org/10.1016/j.cbi.2014.10.016

Masten AS (2007) Resilience in developing systems: progress and promise as the fourth wave rises. Dev Psychopathol 19(3):921–930. https://doi.org/10.1017/S0954579407000442

Masten AS, Obradovic J (2006) Competence and resilience in development. Ann N Y Acad Sci 1094(13):13–27. https://doi.org/10.1196/annals.1376.003

Masten AS, Powell JL (2003) A resilience framework for research, policy and practice. In: Luthar SS (ed) Resilience and vulnerability: adaptation in the context of childhood adversities. Cambridge University Press, Cambridge, pp 1–26. https://doi.org/10.1017/CBO9780511615788.003

McEwen BS (1998) Stress, adaptation, and disease: Allostasis and allostatic load. Ann N Y Acad Sci 1(840):33–44. https://doi.org/10.1111/j.1749-6632.1998.tb09546.x

Narayan D (2000) Voices of the poor: can anyone hear us? (World Bank publication). Oxford University Press, Oxford

Neff LA, Broady EF (2011) Stress resilience in early marriage: can practice make perfect? J Pers Soc Psychol 101(5):1050–1067. https://doi.org/10.1037/a0023809

O'Connell Higgins G (1994) Resilient adults: Overcoming a cruel past. Jossey-Bass Publisher, San Francisco

O'Dougherty Wright M, Masten AS (2005) Resilience processes in development. In: Goldstein S, Brooks RB (eds) Handbook of resilience in childhood. Springer, New York, pp 17–37. https://doi.org/10.1007/0-306-48572-9_2

Ozbay F, Johnson DC, Dimoulas E, Morgan CA, Charney D, Southwick S (2007) Social support and resilience to stress: from neurobiology to clinical practice. Psychiatry (Edgmont) 4(5):35–40

Pearlin LI, Bierman A (2013) Current issues and future directions in research into the stress process. In: Aneshensel CS, Phelan JC, Bierman A (eds) Handbook of the sociology of mental health, 2nd edn. Springer, New York, pp 325–340. https://doi.org/10.1007/978-94-007-4276-5_16

Rutter M (2012) Resilience as s dynamic concept. Dev Psychopathol 24(2):335–344. https://doi.org/10.1017/S0954579412000028

Sedmak C (2013) Innerlichkeit und Kraft. In: Studie über epistemische Resilienz. Herder, Freiburg im Breisgau

Sedmak C (2017) The capacity to be displaced: resilience, mission, and inner strength. Brill, Leiden

Seery MD (2011) Resilience: a silver lining to experiencing adverse life events? Curr Dir Psychol Sci 20(6):390–394. https://doi.org/10.1177/0963721411424740

Sen A (1999) Development as Freedom. Oxford University Press, Oxford

Solomon Z, Berger R, Ginzburg K (2007) Resilience of Israeli body handlers: implications of repressive coping style. Traumatology 13(4):64–74. https://doi.org/10.1177/1534765607312687

Southwick SM, Vythilingam M, Charney DS (2005) The psychobiology of depression and resilience to stress: implications for prevention and treatment. Annu Rev Clin Psychol 1:255–291. https://doi.org/10.1146/annurev.clinpsy.1.102803.143948

Troy AS, Mauss IB (2011) Resilience in the face of stress: emotion regulation as a protective factor. In: Southwick SM, Litz BT, Charney D, Friedman MJ (eds) Resilience and mental health: challenges across the lifespan. Cambridge University Press, Cambridge, pp 30–44. https://doi.org/10.1017/CBO9780511994791.004

Werner E (1992) The children of Kauai: resiliency and recovery in adolescence and adulthood. J Adolesc Health 13:262–268

Werner E (1996) How kids become resilient: observations and cautions. Resiliency in Action 1(1):18–28

Werner EE, Smith R (1982) Vulnerable, but invincible. McGraw-Hill, New York

Werner EE, Smith R (1992) Overcoming the odds: high risk children from birth to adulthood. Cornell UP, Ithaca, NY

Werner EE, Smith R (2001) Journeys from childhood to midlife. Risk resilience and recovery. Columbia UP, Ithaca, NY

Wilks, S. E. (2008). Resilience amid academic stress: the moderating impact of social support among social work students. Adv Soc Work, 9(2), 106–125. 10.18060/51

Wu G, Feder A, Cohen H, Kim JJ, Calderon S, Charney DS, Mathé AA (2013) Understanding resilience. Front Behav Neurosci 7:10. https://doi.org/10.3389/fnbeh.2013.00010

13

Epilogue: The Pandemic as a Big Reveal: Coronavirus, Stress, and Poverty

Any work on stress and poverty after the year 2020 will have a hard time ignoring the pandemic caused by the coronavirus SARS-CoV-2. A crisis is revealing—it unmasks the **vulnerability**[1] of a society. Stress is also revealing—it is a reflection of the interplay of internal and external factors and an indicator of environmental dynamics. A crisis should trigger warning systems and appropriate responses to external pressures such as stress. Members of the European College of Neuropsychopharmacology have "emphasized that **distress** and anxiety are normal reactions to a situation as threatening and unpredictable as the coronavirus pandemic" (Vinkers et al. 2020). Stress is a sign of adaptation, but different bodies or systems are prepared differently for coping with stress. Thus, a crisis is a "stress test" for individuals, but also for the social and political system and for the **resilience** of the infrastructure. The global health crisis caused by the spread of COVID-19 has made the limits of public healthcare systems manifest. It is well understood that the pandemic can be stressful, with its uncertain prospects and unpredictability, with its fatal consequences for hundreds of thousands of people, with its negative impact on economies and livelihoods, with its public consequences such as mobility restrictions and lockdowns, and with its private implications such as social isolation, the experience of disruption, and loneliness. There is now a term for it: "coronavirus stress" (Arslan et al. 2020). Studies that explore the consequences of pandemic-related stress for the brain claim that coronavirus stress can shrink nerve cells and cut their connections (Saunders 2020). There may well be long-term elevated stress levels even after the pandemic, as a study on

---

[1] Glossary terms are bolded at first mention in each chapter.

© Springer Nature Switzerland AG 2021
M. Breitenbach et al., *Stress and Poverty*, https://doi.org/10.1007/978-3-030-77738-8_13

the stress levels of SARS epidemic survivors a year after the 2003 SARS outbreak has demonstrated (Lee et al. 2007).

As Christiaan Vinkers and his coauthors have commented about the
COVID-19 pandemic: "these are stressful times, particularly since the stressor
is new, the absence of warning precluded preparation and pre-adaptation, no
antidotes or vaccinations being currently available, and [there are] unknown
long-term health and society-related implications of the virus. It is unknown
how the pandemic will affect our future lifestyle and when and if we can
resume our regular lives. This pervasive uncertainty makes it difficult to plan
for the future and thus generates additional psychosocial stress" (Vinkers et al.
2020). The additional epistemic labor, i.e., the effort to understand and cat-
egorize the phenomenon, is as stressful as the hard work of living with uncertainty and the unknown. Living with ambiguity and ambivalence is stressful.
The unknown is stressful. Paradoxically, the state of being in suspense—a
kind of fragile equilibrium—is stressful. As an unprecedented phenomenon
in recent times, there are no reference points. It is a situation that could well
lead to "epistemic stress," i.e., the pressure to find appropriate descriptions,
explanations, and strategies, and adequate language to capture what is happening. A pandemic is also stressful in social and economic terms, since it
affects community life and its economic foundations.

The American Psychological Association published the results of a survey
from November 2020 stating that 76% of Americans claim that the pandemic
is a major source of stress (APA 2020). The American Academy of Pediatrics
referred to a study stating that "financial hardships caused by the COVID-19
pandemic are hitting low-income, black and Hispanic families especially
hard" and that "the pandemic exacerbated material hardship and psychological distress for low-income families and warned such stress can have long-
lasting impacts on children" (Jenco 2020).

People who live under chronic stress will experience the acute stress of the
pandemic differently. It is "stress on stress." Acute stress is added to chronic
stress; this means not just increasing the stress levels, but creating a different
quality of the stress situation, a different landscape of stress. After the 2010
earthquake in Haiti, Paul Farmer and colleagues employed the medical expression "acute on chronic." With this phrase they not only described the acute
stress of an earthquake hitting an already severely impoverished population
but also (and especially) a cholera epidemic within a year after the earthquake—a shock health providers were by no means prepared for (Farmer et al.
2011; see also Gabrielli and Lund 2020; Gabrielli et al. 2014).

There is "acute-on-chronic stress" when an acute shock is added to chronic
pressure. There are, however, important differences between an earthquake

and a pandemic: even though an earthquake, as with a pandemic, will have long-term consequences and affect the most vulnerable populations most severely, an earthquake as an event can be more easily predicted and is spatially and temporarily restricted. A pandemic is a much more durable and a much less predictable phenomenon, one that may reach a global scale. Consequently, the stress levels associated with a pandemic are different too. The shock of a pandemic may be less immediate and less tangible, with no building collapsing and no corpses buried under the debris, but the prolonged and extended nature of a pandemic will take its toll and will necessitate **stress responses**.

The pandemic made "stress meet stress," and then, in a similar way, "lockdown meet lockdown," leading to the phenomenon of "a lockdown within a lockdown," which has been the experience in refugee camps. Be it in Bosnia, Greece, or Kenya, the mobility of the refugees in the camps is limited. In some instances, they are not allowed to leave the camp or it is in the middle of nowhere, so they find themselves de facto in a lockdown situation. If the government imposes a lockdown in such a situation, refugees will experience a lockdown within a lockdown. Stress-mitigating factors are internal ("**epistemic resilience**", Sedmak 2013) or social (related to the experience of community and "togetherness"). In 2019/20 Rahul Oka and his research team spent 3 months conducting extensive field research in the Kakuma, Kalobeyei, and Dadaab refugee camps in northern Kenya, which collectively house more than 400,000 refugees. Their research shows "that friendship and faith were consistently associated with lower stress and trauma from the displacement and lockdown, and with higher resilience amongst refugees and hosts" (Oka 2020; see also Oka and Gengo 2020a, b). With its rules on social distancing, the pandemic has made it even more difficult to maintain friendships and the social dimension of faith life, for obvious reasons. The particularly vulnerable have been especially negatively affected by the pandemic.

Vulnerabilities are not evenly distributed. The reciprocal connection between poverty and health (poverty causes health problems, health problems contribute to poverty) is well established; because of the health-related vulnerability of socioeconomically challenged people, it is easily understood that "people who have a history of medical problems and are also suffering from poor health may feel more vulnerable to a new disease" (Salari et al. 2020). Some people have higher life risks, and the risk groups identified during the pandemic are also the ones at higher risk of experiencing stress: older people and people with chronic diseases, children and teens, frontline workers, and people who are helping with the response to COVID-19, such as doctors, other healthcare providers, or first responders. Additionally, people with mental health conditions face higher risks of experiencing stress. Studies have shown "that COVID-19 stress can trigger mild to severe levels of psychosocial problems, such as depression, somatization, and anxiety" (Arslan et al. 2020).

Thus, the pandemic has revealed once again the connection between poverty and stress and its vicious circle: poverty—both material poverty with financial hardships and social poverty with experiences of exclusion and isolation—causes stress, and stress contributes to poverty through lower performance levels, unhealthy response mechanisms, and if stress is chronic, long-term health implications. In short, "chronic stressors inherent with living in poverty, such as food insecurity and reduced access to medical and mental health services, can be exacerbated by the acute stress of COVID-19" (Gabrielli and Lund 2020).

People of low **socioeconomic status** find themselves in stressful, higher-risk situations due to COVID-19: social distancing regulations may increase their feelings of **social exclusion**, which challenge their mental health. In addition, they may not be willing or able to access the healthcare system because of experiences of exclusion or even discrimination, or other barriers, such as transportation or communication. In fact, they may seek medical help later than is ideal and thus experience more serious medical problems, which may negatively impact treatment options. They may live in densely populated areas and in overcrowded accommodations; they are often employed in occupations that do not allow remote work (e.g., cleaning, driving buses, working in supermarkets); if allowed to work remotely, they may face the stress of not having appropriate workspaces, technical resources, or quiet, work-conducive environments; their employment status may be unstable, and they may and face additional levels of anxiety related to potential job loss and loss of livelihood. As medical researchers from London, Liverpool, and Southampton put it: "heightened stress is known to weaken the immune system, increasing susceptibility to a range of diseases and the likelihood of health risk behaviours … Therefore, poverty may not only increase one's exposure to the virus, but also reduce the immune system's ability to combat it" (Patel et al. 2020).

One can use any window to look more deeply into the connection between social protection measures and stress levels. Let us look at the example of the gender dimension of the pandemic; since the majority of frontline care workers are women, a gender analysis of the COVID-19 pandemic may be particularly useful. In many societies, women are more likely than men to be caregivers for the sick and for both children and parents, at home as well as in healthcare settings. Women are more affected by school closures due to the uneven distribution of caring responsibilities, and women have a higher risk of being exposed to domestic violence during quarantines. Women leaving the labor market due to increased domestic responsibilities may suffer long-term economic consequences. Additionally, given educational gaps, women in many countries have less access to information about the pandemic, and women are frequently excluded from a household's major decision-making processes, giving them little control over their own health or that of their children. In short, women are bearing the brunt of the global health crisis (Wenham et al. 2020). Generally speaking, it is clear that the pandemic has exposed women to an increased number of severe **stressors**.

And at the same time—another paradox—the crisis has redefined "essential work" and has declared a number of low-income jobs to be "essential," to the disadvantage of people holding the jobs. An example would be the meat industry. In an interview, Joshua Specht, a historian who described the US meat industry in his *Red Meat Republic*, articulated one concern: "People will feel the pandemic in a much more personal way if they experience shortages." In fact, the US government designated meat-producing facilities "essential" and forced them to stay open, regardless of the risk to workers—more than 15,000 of whom were infected with COVID-19 by late May 2020 (O'Shaughnessy 2020). Here again, many workers are from minority groups and struggling to make ends meet. Similarly, under a general lockdown, the Indian government made the controversial decision to consider the mining industry "essential," probably due to pressure from large companies (Menon and Kohli 2020). Needless to say, working in the meat industry or in mining is stressful because of noise, odors and fumes, crowding, fast pace and pressure, and real danger; in addition, these workers, who exemplify thousands more in a variety of industries, have been labeled "essential," thus, given their working conditions, adding the stress of health risk to the work stress and the societal stress caused by and related to the pandemic.

A further window into the link between stress and poverty is the case of children. Children have fewer cognitive resources to cope with a public health crisis; they are less prepared to understand the concept and background of a pandemic; and they need physical contact and real experiences for proper brain development. Disruption of routines and sudden changes in caregivers can lead to significant levels of stress for children, especially younger children and children with disabilities. Children also suffer from the consequences of parental stress. Sources of stress for parents include the experience of isolation, financial worries, concerns about job security and livelihood stability, disrupted routines, concerns about the health of family members, managing the virtual learning environment of their children, access to appropriate information, and access to adequate healthcare. All these areas of stress hold especially true for vulnerable families and their vulnerable children, for families living in conditions of poverty, for families with high conflict potential, or for families with members facing disabilities. The situation of children and families is another example of where the "acute-on-chronic" stress framework can be applied (Gabrielli and Lund 2020; Patel and Raphael 2020).

Stress is passed from parents to children; parents dealing with high stress levels are impaired in their parenting style and parenting repertoire. In this sense, stress is "transitive" in that it is passed on or transferred within systems and structural hierarchies. Parents who show higher levels of stress during the pandemic create higher stress levels for their children (Spinelli et al. 2020). The well-known **Matthew effect** (advantage begets further advantage: see Rigney 2010) is confirmed in a negative way, as we see so often in poverty research: those who have will be given more, and those who have little will run the risk of losing the little they have.

It is for good reasons that resilience has been identified and emphasized as a key aspect of a proper response to the pandemic (Vinkers et al. 2020), but once again we are confronted with the reality of inequality. In times of quarantine, lockdown, and social isolation, access to digital communications and virtual encounters is important: "mental health professionals recommend promoting healthy behaviors, avoiding exposure to negative news, and using alternative communication methods such as social networks and digital communication platforms to prevent social isolation" (Salari et al. 2020). Experts also emphasize that, "the use of electronic devices and applications to provide counseling can reduce the psychological damages caused by COVID-19, and can consequently promote social stability" (Salari et al. 2020). These are promising strategies, but only for those who have access to the necessary devices. The **digital divide** is a reality in healthy times; it is exacerbated during a pandemic as the need for digital resources increases while the public infrastructure (e.g., schools and public libraries with internet access) decreases or shuts down. It is obvious that a person who is unable to access the internet is confronted with significant barriers to education and entertainment, to virtual communication and virtual community, to online counseling and online medical consultation, to the option of working remotely, to the possibility of accessing information about the pandemic or about state support. Not only are poor people more deeply exposed to pandemic-related stressors, but they are also deprived of important coping mechanisms and strategies to respond to coronavirus stress. The pressure to "do more with less" is one of the sad realities of life for people struggling with poverty and social exclusion.

Let us look at two further risk groups exposed to chronic stress in their daily lives, stress that is hugely and unpredictably increased during the pandemic: the homeless and prison inmates. Homeless people belong to one of society's most vulnerable populations. It is undeniably stressful to live on the streets (Wasserman 2014); there is the stress of being exposed to the elements, the stress of disrespect and insults, the stress of accessing basic resources, the stress of competition on the street, the stress of conflict. This "ordinary stress" of homeless people has been considerably increased by the pandemic. When there is a "stay at home" order, what does this mean for homeless people who have nowhere to go? Many shelters had to close during the pandemic because of risks or restrictions. Margie Pfeil, a theologian and social activist from South Bend, Indiana, worked hard during the pandemic to assist her city's homeless population in finding appropriate shelter during the pandemic. For many weeks, the reality was dire: "Writing from my kitchen table," Pfeil shares, "I look out onto a tent camp of about sixty people here in downtown South Bend. In the midst of COVID-19, there is no indoor shelter available for them, nor are there publicly accessible restrooms anywhere nearby. The only publicly available water is a spigot at the dog park a mile north. Most of the tent residents are African American" (Pfeil 2020). In just a few lines, this quotation addresses living conditions without dignity, fragility, lack of infrastructure and protection, and racial injustice.

Incarcerated people make up a similarly disadvantaged group; again, life in prison is stressful—the stresses of powerlessness, conflict, hierarchy, violence, loss of freedom and control, lack of social status, and separation from family (Skowroński and Talik 2018; Ingram Fogel 1993) are all considerable. Inmates have been exposed to additional stress during the pandemic, since prisons have emerged as high-risk institutions with limited infrastructure for protection from the virus, limited access to healthcare, and widespread phenomena such as overcrowding and limited attention to inmates' health[2]. New health anxieties, further isolation from family members and visitors, and additional regulations are among the stressors the pandemic has added to the already stressful everyday lives of inmates.

There is no doubt that the link between stress and poverty has been exacerbated by the pandemic, whose dynamics expose poor people to higher risk while at the same time reducing their resources for coping. The pandemic has been a poverty trap, with the poverty risks for vulnerable groups increasing significantly—and this rise has happened on a global scale (Diwakar 2020). At the same time, the pandemic is a source of multiple stressors and in this sense is a "stress trap." A crisis like the COVID-19 pandemic reveals social, political, and moral structures by unveiling vulnerabilities and shedding new light on the connection between stress and poverty.

---

[2] There are many studies on this issue, we will mention just one from Africa: Muntingh (2020).

References

APA (2020) Coronavirus pandemic is a significant source of stress. https://www.apa.org/news/press/releases/2020/11/coronavirus-pandemic-stress.   Accessed 11 Jan 2021

Arslan G, Yıldırım M, Tanhan A, Buluş M, Allen K-A (2020) Coronavirus stress, optimism-pessimism, psychological inflexibility, and psychological health: psychometric properties of the coronavirus stress measure. Int J Ment Heal Addict 4:1–17. https://doi.org/10.1007/s11469-020-00337-6

Diwakar V (2020) From pandemics to poverty. Hotspots of vulnerability in times of crisis. London: ODI. https://www.odi.org/publications/16831-pandemics-poverty-hotspots-vulnerability-times-crisis. Accessed 11 Jan 2021

Farmer P, Almazor CP, Bahnsen ET, Barry D, Bazile D, Bloom BR, Bose N, Brewer T, Calderwood SB, Clemens JD, Cravioto A, Eustache E, Jérôme G, Gupta N, Harris JB, Hiatt HH, Holstein C, Hotez PJ, Ivers LC, Kerry VB, Koenig SP, LaRocque RC, Léandre F, Lambert W, Lyon E, Mekalanos JJ, Mukherjee JS, Oswald C, Pape J-W, Prosper AG, Rabinovich R, Raymonville M, Réjouit J-R, Ronan LJ, Rosenberg ML, Ryan ET, Sachs JD, Sack DA, Surena C, Suri AA, Ternier R, Waldor MK, Walton D, Weigel JL (2011) Meeting Cholera's challenge to Haiti and the world: a joint statement on cholera prevention and care. PLoS Negl Trop Dis 5(5):e1145. https://doi.org/10.1371/journal.pntd.0001145

Gabrielli J, Lund E (2020) Acute-on-chronic stress in the time of COVID-19: assessment considerations for vulnerable youth populations. Pediatr Res 88:829–883. https://doi.org/10.1038/s41390-020-1039-7

Gabrielli J, Gill M, Koester LS, Borntrager C (2014) Psychological perspectives on 'acute on chronic' trauma in children: implications of the 2010 earthquake in Haiti. Child Soc 28:438–450. https://doi.org/10.1111/chso.12010

Ingram Fogel C (1993) Hard T: the stressful nature of incarceration for women. Issues Ment Health Nurs 14(4):367–377

Jenco M (2020) Study: COVID-19 pandemic exacerbated hardships for low-income, minority   families.   https://www.aappublications.org/news/2020/06/03/covid19hardships060320. Accessed 11 Jan 2021

Lee AM, Wong JGWS, McAlonan GM, Cheung V, Cheung C, Sham PC, Chu C-M, Wong P-C, Tsang KWT, Chua SE (2007) Stress and psychological distress among SARS survivors 1 year after the outbreak. Can J Psychiatry 52(4):233–240

Menon M, Kohli K (2020) During a lockdown, why is the mining industry considered 'Essential'? https://thewire.in/political-economy/lockdown-mining-steel-essential-regulatory-oversight. Accessed 11 Jan 2021

Muntingh LM (2020) Africa, prisons and COVID-19. J Hum Rights Pract 12:284–292. https://doi.org/10.1093/jhuman/huaa031

O'Shaughnessy B (2020) Slaughterhouse 2.0. Notre dame historian applies research to another hot-button chapter https://www.nd.edu/stories/slaughterhouse-two-point-zero/. Accessed 11 Jan 2021

Oka R (2020) What refugees can teach us about living in crisis https://keough.nd.edu/what-refugees-can-teach-us-about-living-in-crisis/. Accessed 11 Jan 2021

Oka R, Gengo R (2020a) Desk review on resilience building and self-sufficiency among refugees and host communities in CRRF countries. USAID Research Technical Assistance Center, Washington, DC

Oka R, Gengo R (2020b) The political economy of refugee-host integration in Kenya. A comparative case study of barriers to self-sufficiency and resilience in the northern Kenya counties of Turkana and Garissa. USAID Research Technical Assistance Center, Washington, DC

Patel M, Raphael JL (2020) Acute-on-chronic stress in the time of COVID-19: assessment considerations for vulnerable youth populations. Pediatr Res 88:827–828. https://doi.org/10.1038/s41390-020-01166-y

Patel JA, Nielsen FBH, Badiani AA, Assi S, Unadkat VA, Patel B, Ravindrane R, Wardle H (2020) Poverty, inequality and COVID-19: the forgotten vulnerable. Letter to the editor. Public Health 183:110–111. https://doi.org/10.1016/j.puhe.2020.05.006

Pfeil M (2020) The preferential option for the poor and COVID-19. In: Garvey E, Graff D, Gustine A, Hebbeler M, Johnson O'Brien F, Marley Bonnichsen M, Pfeil M, Purcell B, Sedmak C, Watts N, Wilson B (eds) COVID-19 and Catholic social tradition: Reading the signs of the new times. CSC occasional paper series 1/2020. University of Notre Dame, Notre Dame, pp 5–8

Rigney D (2010) The Matthew effect: how advantage begets further advantage. Columbia University Press, New York

Salari N, Hosseinian-Far A, Jalali R, Vaisi-Raygani A, Rasoulpoor S, Mohammadi M, Rasoulpoor S, Khaledi-Paveh B (2020) Prevalence of stress, anxiety, depression among the general population during the COVID-19 pandemic: a systematic review and meta-analysis. Glob Health 16(57):1–11. https://doi.org/10.1186/s12992-020-00589-w

Saunders L (2020) How coronavirus stress may scramble our brains. Imaging studies show we should give ourselves a break. https://www.sciencenews.org/article/coronavirus-covid19-stress-brain. Accessed 11 Jan 2021

Sedmak C (2013) Innerlichkeit und Kraft. Studie in epistemischer Resilienz. Herder, Freiburg/Br

Skowroński BL, Talik E (2018) Coping with stress and the sense of quality of life in inmates of correctional facilities. Psychiatr Pol 52(3):525–542. https://doi.org/10.12740/PP/77901

Spinelli M, Lionetti F, Pastore M, Fasolo M (2020) Parents' stress and Children's psychological problems in families facing the COVID-19 outbreak in Italy. Front Psychol 11:1713. https://doi.org/10.3389/fpsyg.2020.01713

Vinkers CH, van Amelsvoort T, Bisson JI, Branchi I, Cryan JF, Domschke K, Howes OD, Manchia M, Pinto L, de Quervain D, Schmidt MV, van der Wee NJA (2020) Stress resilience during the coronavirus pandemic. Eur Neuropsychopharmacol 35:12–16. https://doi.org/10.1016/j.euroneuro.2020.05.003

Wasserman JA (2014) Stress among the homeless. In: Cockerham WC, Dingwall R, Quah S (eds) Encyclopedia of health, illness, behavior, and society. Wiley-Blackwell Publishers, Oxford, UK. https://doi.org/10.1002/9781118410868.wbehibs327

Wenham C, Smith J, Morgan R (2020) COVID-19: the gendered impacts of the outbreak. The Lancet 395(10227):846–848. https://doi.org/10.1016/S0140-6736(20)30526-2

Glossary

4-Hydroxynonenal (HNE) a highly toxic and mutagenic molecule produced by cells under oxidative stress from unsaturated fatty acids.

8-oxo-deoxyguanosine oxidized form of guanosine in DNA, which is mutagenic and is frequently found in DNA of stressed cells as a typical marker of oxidative stress.

99mTc see *Tc-99m*.

ACTH see *adrenocorticotropic hormone*.

Action potential a transient high (50 millivolts) spike of electrical potential change running down the axon of a nerve cell, thereby transmitting a nerve signal.

Adrenal hormones the hormones produced by the adrenals including epinephrine (a.k.a. adrenaline), cortisol, and aldosterone.

Adrenaline see *epinephrine*.

Adrenal cortex the outer part of the adrenal gland, the place where cortisol and aldosterone are produced.

Adrenal medulla the inner part of the adrenal gland, the place where epinephrine (adrenaline) is produced.

Adrenocorticotropic hormone (ACTH) the peptide hormone produced by the pituitary, which governs hormone production in the adrenal cortex.

Agency the capacity to change situations and exert power based on choices.

Aldosterone the major mineralocorticoid produced in the adrenal cortex; the function of mineralocorticoids is regulation of the sodium/potassium balance by uptake and excretion.

Allostasis a deviation from homeostasis, sometimes also called "cacostasis."

Allostatic load the amount of stressor that disturbs the homeostasis of the organism.

Allostatic overload a state of the organism in which the stressor exceeds the regulatory capacity for a prolonged time, and therefore often leading to onset or aggravation of the physical or mental disease.

© Springer Nature Switzerland AG 2021
M. Breitenbach et al., *Stress and Poverty*, https://doi.org/10.1007/978-3-030-77738-8

AMPA 2-Amino-3-(3-hydroxy-5-methyl-isoxazol-4-yl)propanoic acid; this compound is an agonist of one of the two types of glutamate receptors forming Na^+ channels on the postsynaptic membrane.

Amygdala part of the brain stem just in front of the hypothalamus which regulates emotional behavior.

Aneuploid a cell with an incorrect number of chromosomes.

Anion radical anion radicals are molecules, which are negatively charged (therefore: anions), and at the same time carry an unpaired electron (therefore: radicals); the common name of the oxygen anion radical is superoxide, O_2^-. The dot designates an unpaired electron; the minus designates a negative charge. This molecule can attack proteins and lipids but it is also transformed into hydrogen peroxide, a major signaling substance in cellular metabolism.

Aplastic anemia a disease characterized by loss of regeneration of blood cells.

Apoptosis a process of programmed cell death, which in a multicellular organism can be advantageous, because it removes cells that are already damaged, and if kept, could for example cause cancer. The term is derived from the Greek for "falling of leaves."

Astrocytes a subtype of glia cells of the brain, which is primarily responsible for basic energy metabolism and is in close metabolic interaction with the neurons.

ATP adenosine triphosphate, a nucleoside triphosphate, which is the basic energy transferring molecule of living cells; the major energy-rich compound in cellular metabolism.

Atherosclerosis hardening or stiffening of blood vessels, typically of the arteries in the heart muscle.

Autonomic nervous system consisting of the sympathetic ("activating"), and the parasympathetic ("inactivating") nervous systems; together, these two agents form the autonomic nervous system which, for example, innervates our internal organs, and works subconsciously (autonomically).

Autophagy a process by which the cell destroys and degrades damaged materials, thereby enabling survival; from the Greek for "self-eating."

Biomarkers biochemical changes that occur in cells and organisms under acute or chronic stress; a typical example is 8-oxo-deoxyguanosine. Biomarkers are usually long-lived, and when quantified, measure the accumulation of stressful events over a prolonged time.

BMI see *body mass index.*

Body mass index (BMI) a measure of overweight or obesity based on weight (kilograms) and height (meters); for calculating the BMI, body mass (weight) is divided by the square of the body height and is expressed in units of kg/m^2. Conventionally, a body mass index below 25 (but not lower than 18.5) is considered normal.

Cacostasis see *allostasis.*

Cap a molecular structure at the 5' end of eukaryotic mRNA, which protects mRNA from degradation and is recognized by the eukaryotic protein synthesis machinery.

Catecholamines a class of compounds derived from tyrosine; four hormones belong in this class: L-DOPA, dopamine, norepinephrine, and epinephrine.

CD8 T cells a subtype of T cells in the immune system which is needed for defense against infections and tumors.

Chaperonins/chaperones two types of proteins whose purpose is to help other proteins to fold properly into a three-dimensional structure.

CHD see *coronary heart disease*.

Chromosome puffs Enlarged areas on drosophila chromosomes indicating increased gene expression.

Chronic lymphocytic leukemia a form of cancer characterized by an overproduction of lymphocytes (a subtype of white blood cells).

Complement system a component of the innate immune system that can attack invading microorganisms immediately without prior formation of antibodies.

Coronary heart diseases (CHD) those heart diseases which are caused by stiffening or occlusion of the coronary blood vessels.

Corticotropin-releasing hormone (CRH) a peptide hormone of the hypothalamus which causes the release of corticotropin from the pituitary.

Cortisol the major stress hormone of the adrenal cortex; major functions of this glucocorticoid are: i) stimulation of gluconeogenesis; and, ii) suppression of inflammation reactions of the immune system.

Cortisone oxidation product of cortisol; in humans, cortisone is the normal degradation product of cortisol.

CpG methylation addition of a methyl group to position 5 of the cytosine in the DNA sequence CpG.

CRH see *corticotropin releasing hormone*.

Cytokines a class of immune regulatory peptides causing, among other things, inflammation.

Cytokine storm an acute overproduction of cytokines, which in patients suffering from acute shock can cause multiorgan dysfunction and lead to death.

Cytoplasm the largest part of the intracellular space, containing the protein synthesis machinery and many of the basic metabolic pathways.

Depolarization a change in the neuron's membrane potential induced by the opening of an ion channel and leading (if large enough) to the creation of an action potential.

Dexamethasone a synthetic derivative of cortisol which is used for treating autoimmune and other diseases of a hyperactive immune system.

Differentially methylated region (DMR) a region on DNA, which shows a difference in CpG methylation when stressed and unstressed cells are compared.

Digital divide the gap between those who have access to modern information and communications technology, and those who have no or limited access to these resources.

Disproportionation reaction a chemical reaction whereby two identical molecules are transformed into two nonidentical molecules. To give an example, in the Sod (superoxide dismutase) reaction two ions of superoxide (O_2^-) are transformed into one molecule of O_2 and one molecule of H_2O_2.

Distress large detrimental stress which cannot be responded to adequately.

DMR see *differentially methylated region.*

DNA repair any one of a large number of biochemical processes, which can restore the original unchanged DNA sequence after a chemical insult on DNA.

EGF see *epidermal growth factor.*

EIA see *enzyme immunoassay.*

eIF2alpha eukaryotic initiation factor 2alpha; this protein enables the ribosome to start protein synthesis.

Endemic stress a situation of numerous and continuous—or chronic—stressful demands embedded in everyday life that challenge or even exceed available resources and result in ongoing stressful social conditions.

Endocrine system another term for the hormone system.

Endoplasmic reticulum (ER) one of the organelle systems of a eukaryotic cell; typical reactions taking place in the lumen of the ER are protein glycosylation and protein disulfide formation.

Enzyme immunoassay (EIA) a quantitative procedure of analytical biochemistry for the determination of substances (proteins, small molecules) for which a specific antibody is available.

Eobionts the hypothetical monocellular first living organisms.

Epidermal growth factor (EGF) a protein secreted by epidermal cells, which stimulates the growth of the target cell; stimulation occurs through binding to a surface receptor on the target cell.

Epigenetic mark a chemical change on DNA or chromatin leading to a change in gene expression, which can be transmitted to the next generation.

Epigenetics Transgenerational (intergenerational) inheritance without changes in DNA sequence.

Epinephrine a hormone produced by the adrenal medulla, a.k.a. adrenaline, and a central agent in the stress response; epinephrine is the preferred term in this book.

Epistemic resilience resilience based on inner sources (beliefs, attitudes, memories, faith).

ER see *endoplasmic reticulum.*

Eustasis see *homeostasis.*

Eustress a mild stress that does not kill but on the contrary induces a stress response, which increases the resilience of the cell.

Exhaustion A term used by Selye to describe the final stage of the (see also) General Adaptation Syndrome.

F1, F2, F3, etc. the filial generations 1, 2, 3, etc. in a family pedigree.

GABA see *gamma-aminobutyric acid.*

Gametogenesis the generation of gametes (egg and sperm).

Gamma-aminobutyric acid (GABA) an inhibitory neurotransmitter; this molecule is involved in a long-term depression on certain synapses and in the generation of long-term memory.

GAS see *General Adaptation Syndrome.*

Gene expression the synthesis of products (typically: proteins) following the instructions laid down in DNA sequences; gene expression includes the processes of transcription, translation, and post-synthetic modification. See also: *omics.*

General Adaptation Syndrome (GAS) a term coined by Hans Selye meaning a disease state caused by various kinds of stress or shock and typically leading to diseases—like heart disease or depression—or even death in the long term.

Gentrification the process of gradually changing the character of a neighborhood through the unplanned and planned influx of more affluent residents and businesses.

Germ-line those cells in the body which are destined to form gametes.

Glia cells non-neuronal supporting cells of the brain; among others, two typical functions of glia cells are: energy metabolism of the brain; and formation of the insulating myelin sheath of many neurons which enables and increases neuronal signal transduction.

Glucocorticoid see *cortisol.*

Glucocorticoid receptor a protein that binds glucocorticoids, and after binding of the hormone becomes a transcription factor activating target genes of glucocorticoids.

Glucose-6-phosphate dehydrogenase an enzyme involved in the pentose phosphate pathway, which converts glucose-6-phosphate to 6-phosphogluconate.

Glutathione a non-ribosomally made tripeptide (gamma-glutamylcysteinylglycine), which contains an oxidizable cysteine SH group and is part of the most important redox couple (see *redox couple*) of the cell, often called "redox buffer"; for the structure of reduced and oxidized glutathione, see Fig. <u>3.4</u>.

Glutathione peroxidase gene symbols *GPX1, 2, 3,* etc. (see Fig. <u>3.2</u>); a group of enzymes, which can reduce hydrogen peroxide to water. In the same reaction, reduced glutathione is oxidized.

GSH see *glutathione.*

Haber–Weiss and Fenton reactions these reactions start with superoxide and produce hydroxyl radicals in a cyclic reaction that depends on free Fe^{2+} iron ions (see Fig. <u>3.1</u>b). The hydroxyl radicals are the most reactive and powerful agents of oxidative stress. Therefore, the Haber–Weiss and Fenton reactions are central to the generation of oxidative stress. They can occur in all cells depending on the presence of free ferrous or cuprous ions.

Habitus the physical embodiment of social structures; in other words: appearance, conduct, lifestyle, and taste of a person as shaped by her social position and cultural environment.

Heat shock proteins (Hsps) the proteins produced in a cell as a response to a mild non-lethal heat shock; many of these proteins are chaperones and chaperonins.

Hippocampus part of the brain, which is a central "switchboard" for information processing, in particular for learning and memory; this area of the brain is located on both sides of the medial temporal lobe. It is named for its shape resembling a seahorse.

HNE see *4-hydroxy nonenal.*

Homeostasis a concept quite generally used in biology, which is sometimes also called "eustasis"; it means that the average of the concentration of a certain metabolite (e.g., glutathione) or a certain physical parameter (e.g., body temperature) remains constant, but the actual value undergoes fluctuations. There is a mechanism that restores the set value in the case of environmental fluctuations.

Hormesis a principle of cell physiology by which a mild stress can induce an increased resilience to higher stress levels in a cell, even to stresses not directly related to the primary stress.

HPA axis see *hypothalamic–pituitary–adrenal axis.*

Hsps See *heat shock proteins.*

Hypothalamus part of the brain stem under the thalamus; via releasing hormones, the hypothalamus regulates the activity of the pituitary gland.

Hypothalamic–pituitary–adrenal axis (HPA axis) a hierarchic system of stress hormones produced by the hypothalamus, pituitary, and adrenals; the HPA axis is a major part of the stress response system, which is itself governed by the brain and by hormonal feedback mechanisms.

Identity labor the efforts ("work") necessary to preserve and defend one's identity.

Identity resources the sources that nurture and constitute a person's identity (belonging, recognition, a coherent life narrative, robust concerns).

Idiopathic disease a collective term for disease with an unknown cause.

IGF2 see *insulin-like growth factor2.*

IL-6 see *interleukin-6.*

Imprinting the inactivation of a relatively small number of genes during male and female gametogenesis in mammals.

Innate immunity part of the immune system, which can directly attack invading microorganisms and does not depend on adaptive immunity or the production of antibodies. However, innate immunity and adaptive—or acquired—immunity can interact under conditions of infection.

Insulin-like growth factor2 (IGF2) a protein with sequence similarity to the hormone insulin, which can stimulate the growth of cells; it belongs to a family of proteins of which it is member number 2.

Integrity a moral value and position that reflects a person's consistent and authentic pursuit of a life in accordance with well-justified moral convictions

Intergenerational a trait transmitted from parents to offspring and to all further generations.

Interleukin-6 (IL-6) a protein that acts as a pro-inflammatory cytokine in the immune system

Ischemia a condition characterized by too little blood flow, for instance in the blood vessels of the heart muscle.

Isocaloric nutrition nutrition that contains the same amount of calories as standard nutrition, but may at the same time vary in terms of, e.g., lowered the amount of protein or other important components; such variations are influential, for example, for fetal development.

Lactate the end-product of glycolysis in the absence of cellular respiration.

Language game a definable context of language usage, governed by explicit and implicit rules.

LCC see *leukocyte coping capacity*.

Leukocytes white blood cells; white blood cells are mainly involved in immune defense.

Leukocyte coping capacity (LCC) a parameter for indirectly measuring stress as the inverse of superoxide production by white blood cells in a person exposed to stress; the leukocytes are exposed to Phorbol 12-myristate 13-acetate (PMA), which induces superoxide production. Leukocytes from a stressed person show this reaction in a proportionally reduced amount. The reduction in superoxide is a measure for stress.

Life-world the context of shared everyday practices based on a sense of what is self-evident or given.

Locus coeruleus part of the brain, which regulates the autonomic noradrenergic nerve system.

Longitudinal study a type of clinical or other studies where participants are followed over a prolonged time to investigate the influence of lifestyle factors or medical treatments on the research issue, e.g., disease.

Long-term potentiation (LTP) a process in the hippocampus by which long-term changes in synaptic efficiency are induced in single neurons by priming stimuli; it is believed that this process is part of learning and memory.

Long-term depression (LTD) a persistent decrease in synaptic potential amplitude caused by a primary inactivating signal. Also, this process is essential for memory and learning.

LTP see *long-term potentiation*.

LTD see *long-term depression*.

Major depression a long-lasting disorder of mood consisting of deep melancholy; like stress, major depression activates the HPA axis; in this, major depression and stress seem to result in similar biochemical changes in the body.

Markers of stress biochemical changes, which are easy to quantify, and are used to determine the amount of stress experienced by a person, or of stress present in cells treated with stress.

Matthew effect the dynamics that those in a privileged position will have easier access to more advantages (based on a Gospel verse in Matthew 25:29: "to all those who have, more will be given").

Mental Stress Ischemia Prognosis Study (MIPS) a large international study determining the relationship between psychological stress and the incidence of ischemic heart disease.

Metabolic syndrome a disease of misregulated metabolism, often accompanied by overweight and often preceding diabetes.

Metabolome all the metabolites (small molecular weight substances) that are produced through metabolism in a cell or in an organism.

Membrane attack complex a protein complex formed by the complement system, which perforates the plasma membrane of the target cells and is a central part of the innate immune defense against invading microorganisms.

Mendelian mutation a mutation which is passed on to the next generation according to Mendel's rules.

Mineralocorticoid see *aldosterone*.

MIPS see *Mental Stress Ischemia Prognosis Study*.

Mitochondria (singular: mitochondrion) the organelles of eukaryotic cells, which carry the cellular respiration machinery as well as the machinery for a number of other metabolic reactions.

Mitochondrial diseases diseases caused by mutations in mitochondrial DNA, which lead to severe pathologies, including dysfunction of neurons and of muscles; these diseases are transmitted exclusively from the mother to all of her offspring.

MR brain imaging functional imaging of the live brain by magnetic resonance.

mRNA messenger RNA; this molecule is made from DNA by RNA polymerase and is used by ribosomes in the process of protein biosynthesis.

NADH reduced nicotinamide adenine dinucleotide, a redox-active coenzyme participating in many basic metabolic redox reactions.

NADPH reduced form of nicotinamide adenine dinucleotide phosphate, a redox-active coenzyme; NADPH is needed for all antioxidative reactions in defense of oxidative stress.

NADPH oxidases a class of enzymes that create superoxide starting from oxygen and NADPH. The superoxide produced can serve at least three purposes: (1) antibacterial defense; (2) specialized chemical reactions; and, (3) indirectly creating the signaling substance hydrogen peroxide.

Neuronal plasticity the capacity to change the signaling strength of synapses, and in the long term, to form new synapses.

NMDA see *N-methyl-D-aspartate*.

N-methyl-D-aspartate (NMDA) agonist for one of the two postsynaptic glutamate receptors forming Na^+ channels; this artificial compound increases postsynaptic signaling and depolarization.

NMDA channel an ion channel on the postsynaptic membrane which is sensitive to N-methyl D-aspartate (NMDA). In the presence of NMDA, postsynaptic signaling and depolarization is increased.

Noradrenaline see *norepinephrine*.

Norepinephrine a.k.a. noradrenaline, hormone produced by the adrenal medulla and by dedicated neurons in the sympathetic nervous system; the structure is similar to epinephrine (adrenaline) but lacks one methyl group.

Nox4 NADPH oxidase 4; the human genome encodes seven different NADPH oxidases, Nox4 is one of them. Nox4 is mainly responsible for signaling through hydrogen peroxide.

Omics a collective name for the methods of genomics, transcriptomics, proteomics, and metabolomics—genomics discovers and annotates all genes in a genome; transcriptomics quantifies all gene transcripts in a cell under a certain set of conditions; proteomics does the same for all proteins present in a cell; and metabolomics, finally, for all low molecular weight substances (metabolites). These methods together give a complete picture of gene expression and metabolism in a cell under a given set of conditions.

Oxidative burst a strong but transient production of reactive oxygen species by leukocytes after stimulation by invading microorganisms or pro-inflammatory chemicals.

Oxidative stress state of a living cell with a preponderance of oxidized over reduced biomolecules. Situation when steady-state ROS (reactive oxygen species) concentration is transiently or chronically enhanced, disturbing cellular metabolism and its regulation and damaging cellular constituents.

Pancreatic islet cells the cells of the pancreas, which produce insulin.

Parasympathetic autonomous nervous system see *autonomic nervous system*.

PBMC see *peripheral blood mononuclear cells*.

Pentose phosphate pathway a central glucose metabolic pathway, which is an alternative to glycolysis and produces the largest part of the NADPH needed in the cell.

PEPCK see *phosphoenol pyruvate carboxykinase*.

Perceived Stress Scale (PSS) one of the most prominent instruments for measuring chronic psychological stress; it measures how stressful certain situations are perceived subjectively by a person.

Peripheral blood mononuclear cells (PBMC) a class of blood cells, which plays a major part in innate immunity.

Phorbol 12-myristate 13-acetate (PMA) a pro-carcinogen used in the leukocyte coping capacity stress test.

Phosphoenol pyruvate carboxykinase (PEPCK) a key enzyme of de novo synthesis of glucose (gluconeogenesis).

Phosphoglucose isomerase an enzyme involved in the pathway of glycolysis, which converts glucose-6-phosphate to fructose-6-phosphate.

Pituitary gland an endocrine gland positioned below the hypothalamus in the brain, producing, among other hormones, the adrenocorticotropic hormone.

PMA see *phorbol 12-myristate 13-acetate*.

PNI see *psychoneuroimmunology*.

Post-traumatic stress disorder (PTSD) a psychiatric disease state that can occur in individuals after survival of severe disrupting experiences, like torture or catastrophic events.

Prefrontal cortex part of the brain, which, among other functions, intensively cooperates with the hippocampus in memory formation.

Preimplantation period the very early period of pregnancy before implantation of the embryo into the uterine wall.

Primordial germ cells cells of the early embryo which are destined (pre-programmed) to become gametes.

PSS see *Perceived Stress Scale*.

Psychoneuroimmunology (PNI) a modern field of research in medicine dealing with the interaction of psychological, neurological, and immunological factors.

PTSD see *post-traumatic stress disorder*.

Reactive oxygen species (ROS) a group of biomolecules derived from oxygen (O_2); some ROS are radicals (molecules with unpaired electrons), such as superoxide, and some are non-radicals, such as hydrogen peroxide. ROS are products of the cellular metabolism of oxygen, which may be damaging by attacking DNA, RNA, proteins, and lipids. However, some ROS, for example, hydrogen peroxide, have a biological function because they transmit signals that can be received and can induce changes in cellular metabolism.

Reactive nitrogen species (RNS) biomolecules, which are highly reactive, are derived from nitrogenous compounds and often bear radical electrons on nitrogen; an example is a nitroxyl radical, which is a detrimental radical and at the same time an important signaling substance.

Redox reaction a biochemical reaction in which one of the two partner molecules is oxidized, and the other one is reduced.

Redox couple two molecules which can be interchanged by oxidation/reduction processes; a biologically most important redox couple is that consisting of reduced and oxidized glutathione (Fig. 3.4), which can transiently buffer the oxidative stress inflicted on a cell. Redox couples are involved in many metabolic pathways, like the ones shown in Fig. 3.3.

Resilience the capacity to withstand and survive stress situations and to return to the normal ground state afterward; the ability to cope with a crisis or to flourish in spite of adversity.

Ribosome the organelle of all living cells, which can translate mRNAs into proteins.

Role a set of norms and behavioral expectations based on established beliefs about the position of a person in a social setting.

RNS see *reactive nitrogen species*.

ROS see *reactive oxygen species*.

RONS reactive oxygen and nitrogen species; this is an inclusive term encompassing both kinds of reactive molecules, ROS and RNS.

SA beta-galactosidase senescence-associated beta-galactosidase; an enzyme, which is found in senescent cells, and that is considered as a marker indicating senescence.

Self-efficacy the conviction that one can change the world through one's own efforts and choices.

Serotonin reuptake drugs a class of antidepressant drugs, which block the reuptake of serotonin in the presynaptic cell, based on the notion that in many (but not all) patients depression is caused by a deficiency in serotonin metabolism. A well-known example of serotonin reuptake drugs is Prozac.

SES see *socioeconomic status*.

Single-electron reduction steps of molecular oxygen the addition of a single electron to molecular oxygen creates superoxide; superoxide can be further metabolized to many ROS (see Fig. 3.1a).

Social exclusion the inability to participate in standardized cultural activities against the will of the excluded person or group.

Social capital the total of relations between individuals or groups based on mutual recognition, shared norms, and values; social capital can contribute to mutual benefit and advantageous cooperation, hence to achieving an individual as well as collective good. While "bonding" social capital describes relationships within a rather homogenous group, "bridging" social capital can help a person to move beyond her established social sphere.

Social Readjustment Rating Scale (SRRS) a method employing questionnaires to determine the occurrence and intensity of major change-requiring life events; the stress-values for each item on this scale are predefined, they were initially empirically assigned by a pilot group consisting of nearly 400 participants.

Socioeconomic status (SES) a measure based on income, educational status, and employment status; "low socioeconomic status" (low SES)—i.e., low income, low educational level, and precarious employment/unemployment—is one possible way of conceptualizing poverty and, as data is rather easily available (compared to other concepts) widely used in research.

SOD see *superoxide dismutase*.

SSRI selective serotonin reuptake inhibitor; a common class of antidepressant drugs. See also: *serotonin reuptake drugs*.

SRRS see *Social Readjustment Rating Scale*.

Standardized mean change a measure of stress which adjusts for individual differences between the probands.

Stressor the initial cause for an organism's stress response; this can be, e.g., an external stimulus, a biochemical agent or substance, environmental or social conditions, or life events and changes. See also: *distress*, and *eustress*.

Stress response all changes of the metabolism of a cell or organism under the influence of stress; this may be accomplished at all levels of gene expression, including transcription, translation, degradation of gene products, and at the level of enzyme activity. Stress responses generally ensure the survival of the cell, but may also lead to apoptosis (programmed cell death).

Superoxide dismutase (SOD) an enzyme converting superoxide to oxygen and hydrogen peroxide; technically, such a reaction is called a dismutation. It serves to detoxify the superoxide radical, and in some cases to create the signaling substance hydrogen peroxide.

Sympathetic autonomous nervous system see *autonomic nervous system.*

Takotsubo syndrome a non-ischemic heart disease, which is primarily caused by psychological stress and results in an enlarged left ventricle of the heart. The name of the syndrome stems from the shape of the enlarged left ventricle, which is evocative of the traditional Japanese octopus traps, called Takotsubo. Another name is "broken heart syndrome," which emphasizes psychological stress as a cause for this disease.

Tc-99m a short-lived radioisotope of the element technetium which is used in routine diagnosis of ischemic heart disease; another way to designate the isotope is ^{99m}Tc.

Telomerase an enzyme, which can elongate the telomeres (ends) of chromosomes.

Telomeres the ends of the linear chromosomes of higher organisms.

Thalamus part of the brain stem, which relays sensory information to the cerebral cortex.

Thioredoxin pathway gene symbol *TRX1, 2* (see Fig. 3.2); this pathway can reduce alkyl hydroperoxides to the respective alcohols and is part of antioxidative detoxification.

Toxic stress a repeated or chronic stress condition that results in a chronically activated stress response without rest and recovery and can thus lead to detrimental health outcomes, both physically and mentally.

Translation the process of protein synthesis on the ribosome.

Transgenerational inheritance a type of inheritance that is passed from one generation to the next. The meaning of "transgenerational" is similar to "intergenerational."

Trier Social Stress Test (TSST) a stress test using laboratory situations, such as speaking in front of an audience, which can be used to investigate the correlation between perceived psychological stress and biochemical measurements of stress, mainly stress hormones.

Trier Inventory for the Assessment of Chronic Stress (TICS) a method for assessing area-specific chronic stress, e.g., in occupational or social settings.

TSST see *Trier Social Stress Test.*

TICS see *Trier Inventory for the Assessment of Chronic Stress.*

Vulnerability the exposure to risks with limited or lacking coping resources; vulnerability can be understood as the opposite of resilience.

Printed in the United States
by Baker & Taylor Publisher Services